Everyday Life is Full of Math

This book explores how mathematics appears in everyday life. It presents math in a fun and beautiful way using knowledge at the junior high and high school level. It is written for general readers, not for experts. The book avoids difficult math and instead focuses on how new ways of thinking and careful observation can reveal interesting math ideas. It is divided into 33 topics. These include familiar recreational math like tournament games, magic squares, and math tricks, as well as unique ideas like the geometry of origami and toy train tracks. Some of the topics in this book may be unfamiliar to readers outside Japan, as they are based on things the author has observed in daily life in Japan. However, this is also one of the unique features of the book. It offers readers a glimpse into everyday life in Japan. The goal is to help readers feel closer to math, rather than to provide deep academic content.

Features

- Easy to understand with junior high or high school level math knowledge.
- Introduces math found in real life, like origami.
- Includes many topics based on the author's popular social media posts (over 23,000 followers on X).
- Connects math topics with hands-on activities and experiences.
- Helps readers see the world in a new way through a "mathematical way of thinking".
- Focuses on intuitive and visual understanding, not difficult formulas or theories.
- Provide links (with QR codes) to puzzle apps developed by the author.

AK Peters/CRC Recreational Mathematics Series

Series Editors

Robert Fathauer
Snezana Lawrence
Jun Mitani
Colm Mulcahy
Peter Winkler
Carolyn Yackel

Parabolic Problems
60 Years of Mathematical Puzzles in Parabola
David Angell and Thomas Britz

Mathematical Puzzles
Revised Edition
Peter Winkler

Mathematics of Tabletop Games
Aaron Montgomery

Puzzle and Proof
A Decade of Problems from the Utah Math Olympiad
Samuel Dittmer, Hiram Golze, Grant Molnar, and Caleb Stanford

A Stitch in Line
Mathematics and One-Stitch Sashiko
Katherine Seaton

The Four Corners of Mathematics
A Brief History, from Pythagoras to Perelman
Thomas Waters

Intermediate Poker Mathematics
Mark Bollman

Mathematical Meditations
Snezana Lawrence

The Secret World of Flexagons
Fascinating Folded Paper Puzzles
Scott Sherman, Yossi Elran, and Ann Schwartz

The Magic Theorem
A Greatly-Expanded, Much-Abridged Edition of The Symmetries of Things
John H. Conway, Heidi Burgiel and Chaim Goodman-Strauss

Everyday Life is Full of Math
Jun Mitani

For more information about this series please visit: https://www.routledge.com/AK-Pe-tersCRC-Recreational-Mathematics-Series/book-series/RECMATH?pd=published,forth coming&pg=2&pp=12&so=pub&view=list

Everyday Life is Full of Math

Jun Mitani

CRC Press
Taylor & Francis Group
Boca Raton London New York

CRC Press is an imprint of the
Taylor & Francis Group, an **informa** business

AN A K PETERS BOOK

First edition published 2026
by CRC Press
2385 NW Executive Center Drive, Suite 320, Boca Raton FL 33431

and by CRC Press
4 Park Square, Milton Park, Abingdon, Oxon, OX14 4RN

CRC Press is an imprint of Taylor & Francis Group, LLC

© 2026 Jun Mitani

ISBN: 978-1-041-13538-8 (hbk)
ISBN: 978-1-041-13349-0 (pbk)
ISBN: 978-1-003-67026-1 (ebk)

DOI: 10.1201/9781003670261

Typeset in CMR10 font
by KnowledgeWorks Global Ltd.

Publisher's note: This book has been prepared from camera-ready copy provided by the authors.

Contents

Preface

Our everyday lives are full of things related to mathematics. Considering that mathematics has developed alongside our efforts to understand the natural world, this is only natural.

What kind of image have you had of "mathematics" up until now? Rows of numbers and symbols in textbooks, or complicated equations, might seem far removed from our daily lives. But the things we've learned in school actually help us make sense of the world around us. If you observe carefully, you'll find mathematics hidden in many aspects of daily life. The process of deepening your understanding through math is full of small discoveries and fresh surprises—it's a truly enjoyable experience. You don't need advanced knowledge; just a slight shift in perspective is enough to make sense of things using the math you learned in middle and high school.

That said, changing the way we look at things isn't always easy. That's why in this book, I've gathered 33 fun and easy-to-understand topics that explore mathematics in everyday life, based on what I've discovered and thought about. The topics jump around quite a bit, and it might feel a bit unstructured, but that's because the math I've encountered comes from all corners of daily life. There isn't a single storyline to "everyday math"—it appears suddenly and unexpectedly. Through this book, I hope to share with you the moments where my daily life intersected with mathematics and the enjoyment I found in those moments.

By reliving the joy I experienced, I hope you'll gain a new "mathematical lens" for looking at the world around you. With it, even the smallest moments in daily life might turn into mathematical discoveries.

I've loved making things ever since I was a child, so many of the topics in this book are related to crafting and creating. I encourage you to try these out for yourself. There's a lot of mathematics in making things. As you work with your hands and bring ideas to life, you'll likely feel the fun and wonder of mathematics even more clearly.

Many of the topics in this book were originally shared on X (formerly Twitter), where I regularly post thoughts and observations from everyday life. Here, I've expanded on what couldn't be fully expressed in short posts, offering a wide range of topics.

Through this book, I hope you'll come to enjoy the mathematics hidden in your daily life.

Author Bios

Jun Mitani is a professor of Information and Systems at the University of Tsukuba. He received his Ph.D. in engineering from the University of Tokyo in 2004. He has been present post since April 2015. His research interests center on computer graphics, particularly geometric modeling techniques and their application to origami design. The origami artworks created by him have features that are three-dimensional shapes with smooth curved surfaces. His main books are "3D Origami Art (2016)" and "Curved-Folding Origami Design (2019)". In 2010, through an exchange with ISSEY MIYAKE, he contributed to the launch of the new 132.5 fashion brand. He also cooperated in the design of origami used in the movies "Shin Godzilla (2016)" and "Death Note Light up the NEW world (2016)". His unique origami has been well received around the world, and he had received invitations to hold workshops and exhibitions in Germany, Switzerland, Italy, Israel and many other countries. His work had inspired the design of the trophy for the Player of the Match winner of each game at the Rugby World Cup 2019. His major awards are "Microsoft Research Japan Informatics Research Award, 2012" and "The 2nd Japan Society for Graphic Science Award, 2007". He was appointed as a Japan Cultural Envoy from the Agency for Cultural Affairs and traveled to eight Asian countries to promote cultural exchanges through origami in 2019.

Contributors

Koji Yoshiike
Illustrator

I

The Wonders of Numbers and Shapes Around Us

Even the most ordinary aspects of daily life, when viewed through the lens of "mathematics," are filled with fresh surprises and discoveries. Let's enjoy finding patterns and principles in familiar topics and thinking about them mathematically.

Is "Paper" the Ultimate Strategy in a Rock-Paper-Scissors Tournament!?

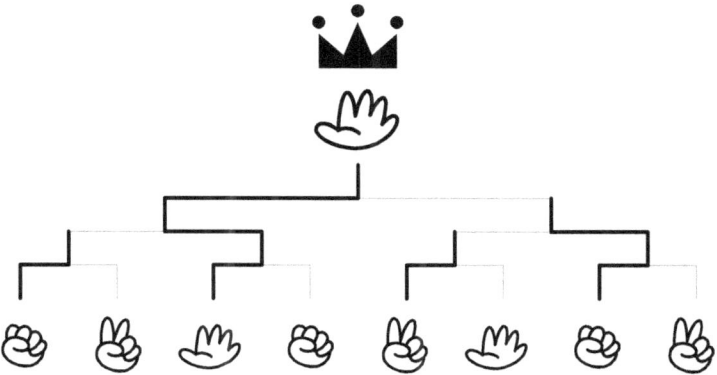

Figure 1.1 Rock-paper-scissors tournament bracket.

What happens if we hold a tournament using rock, scissors, and paper? In a tournament format, there's always a single winner, so we should be able to determine the strongest move. For a start, let's try arranging 8 players in the order of rock, scissors, and paper, and run the tournament.

DOI: 10.1201/9781003670261-1

As shown in the diagram above, "paper" ends up winning. So, does that mean paper is the strongest?

TOURNAMENT FORMAT

The tournament format is a widely used method for determining a winner among multiple participants, commonly seen in various sports competitions. In the first round, half of the participants are eliminated, and only the remaining half advance to the next match. In the second round, again, half are eliminated. In this way, the format efficiently (and perhaps mercilessly?) narrows down the field by continuously removing the losers from the competition. With each round, the number of remaining contestants is halved until finally, the last two face off in the final match to determine the champion.

NUMBER OF MATCHES TO WIN A TOURNAMENT

Let's think about the relationship between the number of matches one must win in a tournament and the number of participants.

If there are 2 participants, winning 1 match secures the championship. With 4 participants, it takes 2 wins to become the champion. With 8 participants, 3 wins are needed. As the number of participants doubles, the number of matches required to win the tournament increases by one.

This relationship can be expressed with the following formula:

$$\text{Number of matches} = \log_2(\text{Number of participants})$$

Here, the symbol $\log_2$ appears.[1] This notation represents "to what power 2 must be raised to get the number of participants," and that power corresponds to the number of matches needed to win.

It might be easier to understand the relationship if we write it as:

$$2^{\text{Number of matches}} = \text{Number of participants}$$

This expression conveys the same idea as the earlier formula, just written in a different way.

However, this simple relationship only holds when the number of participants is a power of 2—such as 2, 4, 8, 16, or 32.

---

[1]$\log_2 n$ means "to what power must 2 be raised to equal n." For example, when $n = 16$, $\log_2 n = 4$, because 16 is equal to 2^4.

For example, if there are 7 participants, one person will have to skip the first round and join from the second round instead. This is known as a "bye" or a "seed."[2]

In such cases, as shown in the diagram below, the number of matches needed to win varies depending on the participant, making the tournament structure seem less fair.

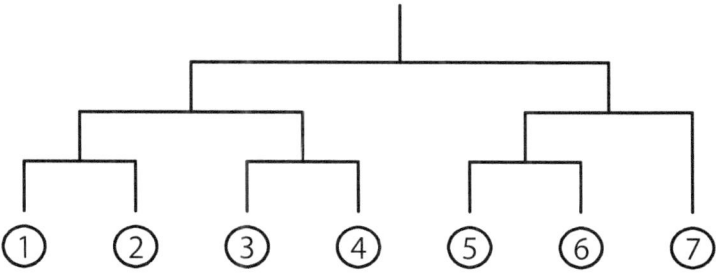

Figure 1.2 Tournament held among 7 people.

TOTAL NUMBER OF MATCHES IN THE TOURNAMENT

In a tournament, each match results in one participant being eliminated. This means that with each match, the number of remaining players decreases by one until only the winner remains undefeated. Therefore, if a tournament starts with n participants, the total number of matches held will be $n - 1$. This rule holds true even when there are seeded players.

For example, in a tournament with 7 participants, there will be 6 matches in total (you can count them to confirm).

In a "fair" tournament, where the number of participants n is a power of 2, the number of matches in each round decreases by half in each subsequent round. In the first round, there are $n/2$ matches, in the second round $n/4$, and so on.

Since the total number of matches is $n-1$, we arrive at an interesting equation.

---

[2]The term "seed" comes from the idea of "scattering seeds," and it refers to the practice of placing strong players strategically throughout the tournament bracket. This is done to prevent top contenders from facing each other in the early rounds.

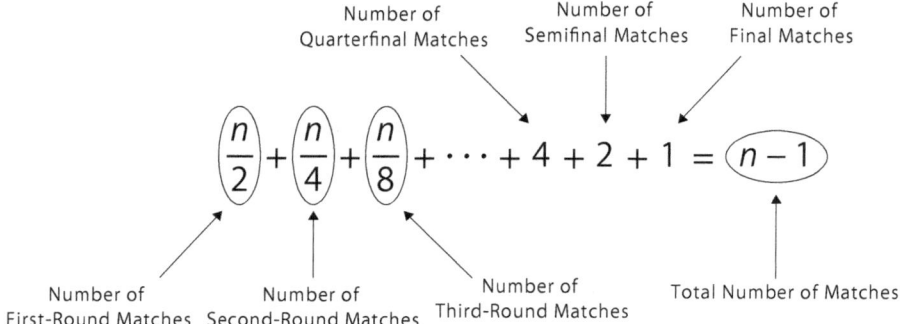

When n is a power of 2, and we keep halving the number until we reach 1, the sum of all those halved numbers is exactly $n - 1$.

For example, when $n = 32$,

$$16 + 8 + 4 + 2 + 1 = 31,$$

which indeed equals $n - 1$.

ROCK-PAPER-SCISSORS AND ROUND-ROBIN TOURNAMENTS

Now, let's turn our attention to rock-paper-scissors. In this familiar game, rock beats scissors, scissors beats paper, and paper beats rock. If we try to arrange the hands in order of strength, we end up going around in a circle, as shown in the diagram below, and can't determine a single strongest hand.

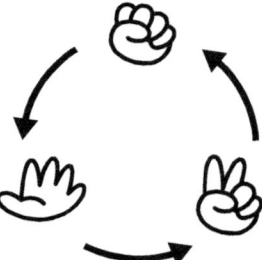

Figure 1.3 Rock-paper-scissors relationships (arrows point the stronger hand).

This kind of relationship is called a *three-way deadlock* or *three-way standoff*. If rock, paper, and scissors were to compete in a *round-robin tournament* (where everyone plays against everyone else), each hand would end up with 1 win and 1 loss. This shows that all three are equally strong. (The number of matches in a round-robin tournament with n participants is given by $\frac{n(n-1)}{2}$.)

Table 1.1   Round-Robin Results of Rock-Paper-Scissors

	✊	✌	🖐	Result
✊		Win	Loss	1 Win, 1 Loss
✌	Loss		Win	1 Win, 1 Loss
🖐	Win	Loss		1 Win, 1 Loss

ROCK-PAPER-SCISSORS AND TOURNAMENT-STYLE MATCHES

So, what happens if rock, paper, and scissors compete in a tournament-style match? In a tournament, there must always be exactly one winner, so it seems like we should be able to determine the strongest hand.

However, with only three hands—rock, paper, and scissors—it's impossible to create a fair tournament bracket. One of the hands will have to be given a bye (a seed).

As a result, as shown in the diagram below, the hand that receives the bye ends up winning the tournament.

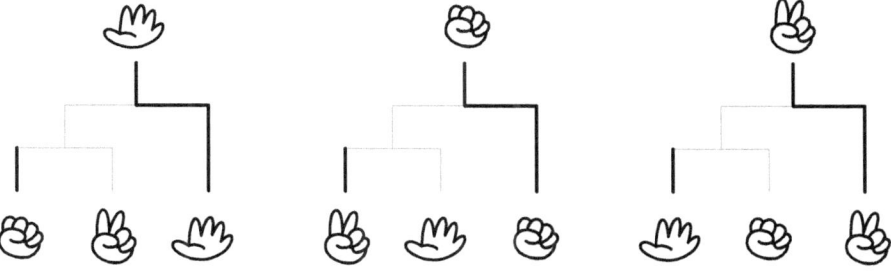

Figure 1.4   Rock-paper-scissors tournament with a seeded hand.

It's interesting to think about why the hand that receives the bye ends up winning.

In this setup, the winner of the first match is guaranteed to lose in the next round. If the winner of the first match were to win again, it would mean that this hand beats both of the other two—which contradicts the nature of the three-way deadlock.

That's why the hand that enters in the second round (the one given a bye) ends up winning.

Even though all three hands are equally strong, the outcome depends entirely on how the matchups are arranged. This shows that the tournament format itself may not be fair in situations like this.

To make a fair tournament bracket, let's add an extra "rock" after the usual rock-paper-scissors sequence, making it a four-participant tournament. This means that rock will participate twice, but that's a compromise we have to accept.

In this setup, as shown in the diagram below, paper ends up winning the tournament.

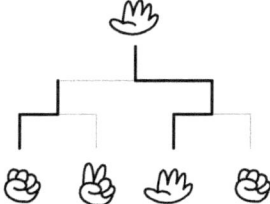

Figure 1.5 Rock-paper-scissors tournament with 4 participants.

This time, let's double the size of the tournament bracket and create a tournament with 8 participants.

Again, we'll arrange the hands in the order rock, scissors, and paper from left to right.

This is the same tournament diagram that was shown at the beginning.

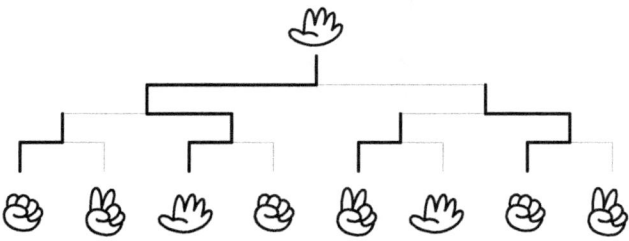

Figure 1.6 Rock-paper-scissors tournament with 8 participants (same as the Figure 1.1).

Once again, paper wins the tournament! Does paper always win in a tournament-style match like this?

Let's try doubling the size of the tournament bracket once more to find out.

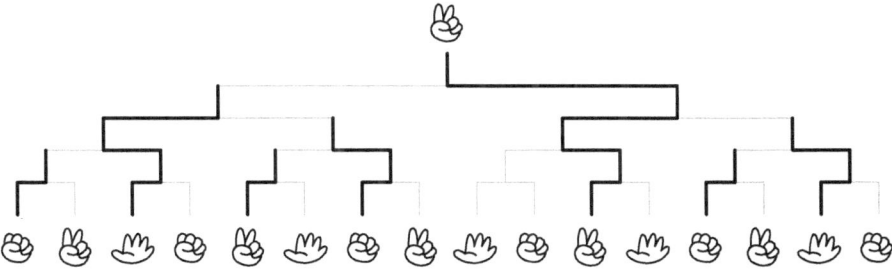

Figure 1.7 Rock-paper-scissors tournament with 16 participants.

This time, scissors won the tournament. How curious!

Starting with 4 participants and repeatedly doubling the number, the tournament winners change as follows.

🖐 → 🖐 → ✌ → ✌ → ✊ → ✊ → 🖐 → 🖐 → ✌ → ✌

→ ✊ → ✊ ···

It's quite unexpected to discover such a pattern!

Now, let's assume the tournament bracket is infinitely large. In that case, what was initially lined up from the left as:

✊ → ✌ → 🖐 → ✊ → ✌ → 🖐 → ···

will, after the first round is completed, become the following order:

✊ → 🖐 → ✌ → ✊ → 🖐 → ✌ → ···

After the second round is completed, the order will change to:

🖐 → ✊ → ✌ → 🖐 → ✊ → ✌ → ···

(Let's try it out).

If we represent Rock, Paper, and Scissors with the numbers 1, 2, and 3 respectively, the arrangement that was initially 123... will, as the matches progress, go through 6 changes in order as follows, before returning to the initial state:

$$123 \to 132 \to 312 \to 321 \to 231 \to 213 \to 123$$

In any of these arrangements, the same number never appears next to itself, which means there are no ties during the matches. Incidentally, there are only 6 possible arrangements for three elements, and we can confirm that all 6 of these arrangements appear once. It's very interesting, isn't it?

After 6 rounds are completed, we can confirm that the 123... arrangement, that is

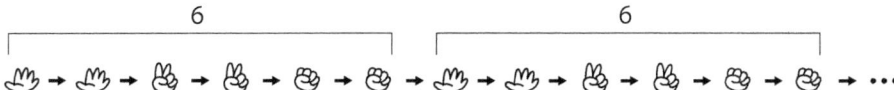

which is the initial state, reappears. The reason why the winner cycles through in 6 rounds, as shown in the diagram below, when the number of tournament participants is doubled and doubled again, is due to this principle.

I encourage all of you to try drawing out your own tournament brackets and experimenting with different setups.

At first glance, a rock-paper-scissors tournament might not seem very interesting—but as we've seen, it can lead to some surprising and fascinating discoveries.

The discovery here leads to the concept of a *group*, which is studied in university-level mathematics. However, without going that far, let's bring this chapter to a close.

CHAPTER 2

Math Magic with Tricks and Twists

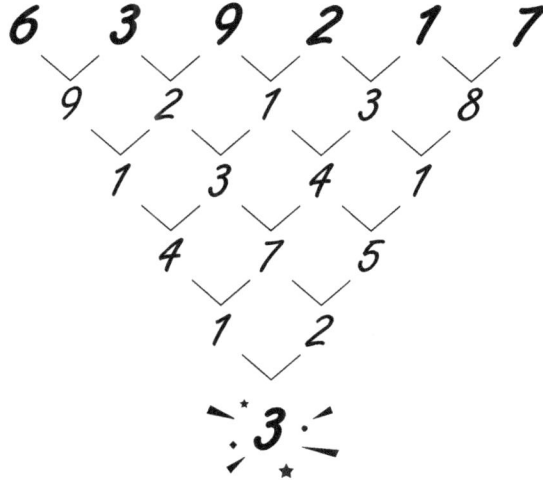

Figure 2.1   Derive the bottom number from the six numbers above.

There is a fun little math trick that can amaze the people around you with just a bit of calculation. Of course, there's a hidden mechanism behind it, but anyone seeing it for the first time is sure to be surprised.

DOI: 10.1201/9781003670261-2

GUESSING THE FINAL NUMBER THAT COMES OUT AT THE END

I once learned a very interesting math trick from a math teacher at a drinking party. It was quite a while ago, so I've forgotten the teacher's name, but the trick itself was so amusing that I still remember it clearly. Let me share it with you here.

First, ask someone to write down six single-digit numbers of their choice in a row. For example, let's say the numbers are 6, 3, 9, 2, 1, and 7, as shown in the illustration at the beginning.

Next, follow these steps to ultimately arrive at a single-digit number (it may be easier to understand by looking at the illustrations than by reading the following explanation):

1. Look at the numbers from left to right and add each pair of adjacent digits.

2. Write down the last digit of each sum (for example, $5 + 7 = 12$, so you write down 2).

3. Repeat this for all six numbers to create five new numbers.

4. Perform the same operation on these five numbers to create a new set of four numbers.

5. Continue this process, and you will eventually end up with a single digit.

Now comes the magic. You can guess what that final digit will be the moment the six original numbers are written down. As soon as the person finishes writing the six digits, you can boldly say, "The result will be 3," or write it down beforehand and reveal it later to surprise everyone.

Even though it takes quite a bit of effort to manually derive the answer, you're able to announce it instantly—now that's impressive!

Now, let me explain how you can guess the answer.

Assign numbers ① through ⑥ to the six digits written on the paper from left to right. The final digit will be determined as follows:

- If ② + ⑤ is even: take the last digit of ① + ⑥.

- If ② + ⑤ is odd: take the last digit of ① + ⑥ + 5.

In the example shown at the beginning, the second number from the left (3) and the fifth number (1) add up to an even number. So, we add the first number (6) and the last number (7), which gives 13. The last digit of that is "3," and that's the answer.

Not too hard, right? Once you memorize this trick, you'll be able to amaze everyone around you!

THE SECRET BEHIND THE MATH TRICK

Let's reveal the secret behind how we're able to guess the final number using the previous method.

Suppose we represent the six numbers lined up from the left as ①, ②, ③, ④, ⑤, and ⑥. We'll go ahead and perform the additions step by step until we're left with just one final number.

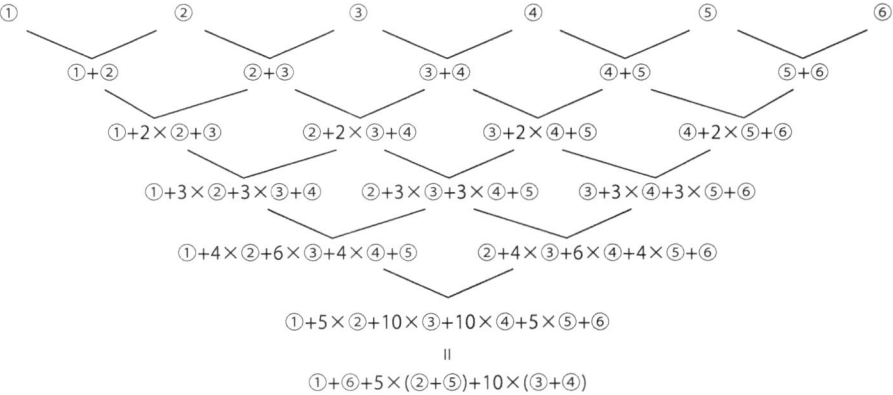

Figure 2.2 Calculation process.

Then, as shown in the figure above, the final result becomes:

$$① + 5 × ② + 10 × ③ + 10 × ④ + 5 × ⑤ + ⑥$$

Rewriting this expression, we get:

$$① + ⑥ + 5 × (② + ⑤) + 10 × (③ + ④)$$

Now, the term $10 × (③ + ④)$ always ends in 0, no matter what values ③ and ④ are, so it does not affect the final digit of the result.

When ② + ⑤ is even, $5 \times (② + ⑤)$ becomes a multiple of 10, so its last digit is also 0. As a result, the final digit is simply the last digit of ① + ⑥.

On the other hand, when ② + ⑤ is odd, we can write it as $2n + 1$, so:

$$5 \times (② + ⑤) = 5 \times (2n + 1) = 10n + 5$$

In this case, the last digit is 5, and the final result becomes the last digit of ① + ⑥ + 5.

And that's the secret behind the trick!

Now, why not try this trick right away with your family or friends? And here's a tip to make it even more impressive.

When deciding on the initial six numbers, subtly guide the process so that ② + ⑤ turns out to be an even number. If you do the trick a few times this way, some sharp observers might catch on and say, "You just add ① and ⑥, right?"

At that point, switch things up and create a case where ② + ⑤ is odd. This will surprise them—"Wait, that's not the answer?!"—and it'll make it much harder for them to figure out the trick.

CHAPTER 3

The Numbers Around Us: Which Digit is Used the Most?

Figure 3.1   Prices of various products listed in supermarket flyers.

Supermarket flyers, which often come with newspapers, are now increasingly available online instead of in print. They list prices for a variety of products, with lots of numbers appearing throughout. If we were to count how many times each digit from 0 to 9 appears, might we discover something interesting?

NUMBERS USED IN SUPERMARKET FLYERS

Even supermarket flyers reveal interesting discoveries when you take a closer look. When you see the prices of products listed in a flyer,

DOI: 10.1201/9781003670261-3

you might often feel like they're "a bargain," right? So why not take a moment to consider what kinds of numbers are actually being used?

For example, you may notice that many of the prices end with the digit 8. This is a psychological trick that makes people feel like they're getting a good deal—a kind of numerical magic.

So, let's collect all the product prices listed in a supermarket flyer and examine how often each digit from 0 to 9 appears as the last digit. What were the results? Here's what I found when I tested it using a flyer I had at home:[1]

Table 3.1 Frequency of Digits 0–9 in Supermarket Flyer Prices

Digit	0	1	2	3	4	5	6	7	8	9
Frequency	15	11	8	7	6	5	4	8	19	6

The prediction that the digit 8 would appear most frequently was spot on. On the other hand, the digit that appeared the least was 6, only about one-fifth as often as 8. Indeed, prices ending in 6 are rarely seen.

Also, most products sold in supermarkets are priced in the range of 100 to 500 yen, so there aren't many items priced in the 600-yen range.[2] This likely explains why the digit 6 showed up so infrequently.

Now, following 8, the next most frequent digit was 0. At first glance, 0 might seem to be at a disadvantage since it can't appear as the first digit of a price. However, it likely ranked high because many prices end in 0, such as 120 yen or 180 yen.

When we plot the frequency of each digit in a graph, the result looks like the figure below.

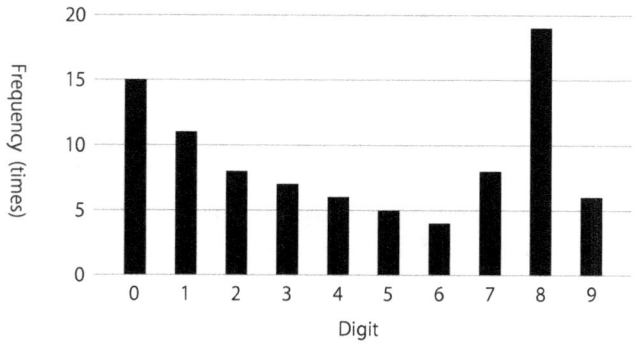

Figure 3.2 Frequency of digits 0–9 in supermarket flyer prices.

---

[1] I live in Japan.

[2] At the time of writing, 100 yen is approximately 0.7 dollars.

Since the sample size was small, this pattern might just be a co-incidence—but it's still interesting to see the numbers decrease from 0 to 6.

This time we looked at supermarket flyers, but if we examined advertisements for high-priced items like cars, the digit 0 might appear even more frequently. What about digits in phone numbers, apartment room numbers, or calendar dates? Investigating the distribution of digits from 0 to 9 in the numbers around us might lead to some fascinating discoveries.

BENFORD'S LAW

The distribution of digits we looked at earlier was based on the prices of items listed in a supermarket flyer. In other words, these numbers were deliberately chosen to appear attractive to customers considering a purchase. Because of that, it's not guaranteed that numbers from other contexts will follow the same pattern.

Now, let me introduce an interesting idea called *Benford's Law*. This law states that if you focus on just the first digit of various natural numbers—like the populations of prefectures and municipalities, stock prices, lengths of rivers, or heights of mountains—the distribution isn't uniform. Instead, the number "1" appears most frequently. Is that really true?

Let's try checking this by looking at the population of Japanese municipalities. We'll examine the first digit of each number in the statistical data published by the Japan's government, specifically from the 2023 Basic Resident Register. Your own country probably has a wide range of government-published data too. Just browsing through what's available can be quite fun.

Now, taking a look at the numbers in the population data for Japan's prefectures and municipalities, the largest number is the population of Tokyo, which is 12,548,258, and the smallest is Aogashima Village, with a population of 157 (both of which start with the digit "1"). It might seem strange to line up the population of Tokyo with that of a small village, but since we're interested in the numbers themselves, let's go with it. In total, we examined 2,336 numbers.

Next, let's take a look at the distribution of the first digits of these numbers. Since the first digit of a number can't be 0, we'll focus on how often the digits from 1 to 9 appear.

What do you think the result will be?

Here's what we found.

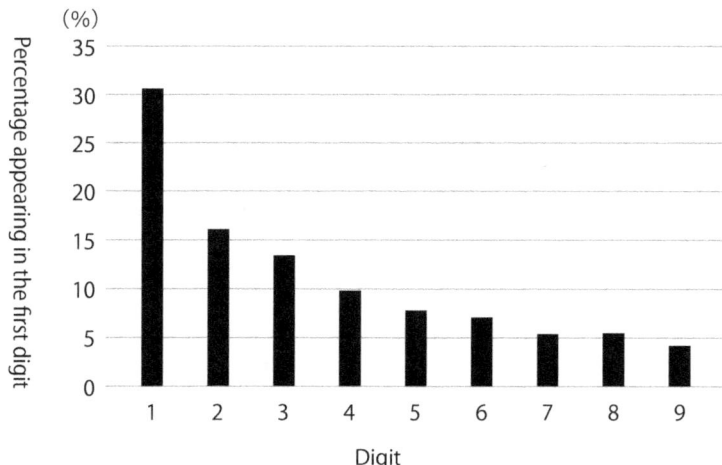

Figure 3.3 Distribution of the first digit of the population by prefecture and municipality.

We can see that about 30% of the numbers start with the digit "1," which is overwhelmingly more than any other digit. This is quite interesting, especially when you compare it to the distribution of digits found in supermarket flyers—the patterns are completely different.

For example, if values are evenly spread across the range from 500 to 1500, about half of them would naturally start with a "1." So, even intuitively, it makes some sense that "1" might appear more often.

According to Benford's Law, the probability $P(d)$ that the first digit of a number is the digit d is given by the following formula:

$$P(d) = \log_{10}\left(\frac{d+1}{d}\right)$$

(A brief explanation of the $\log_{10}$ notation will be provided in the column on the next page.)

When we apply this formula to the digits from 1 to 9 and compare the calculated values with the results we observed, we get a graph like the one shown below.

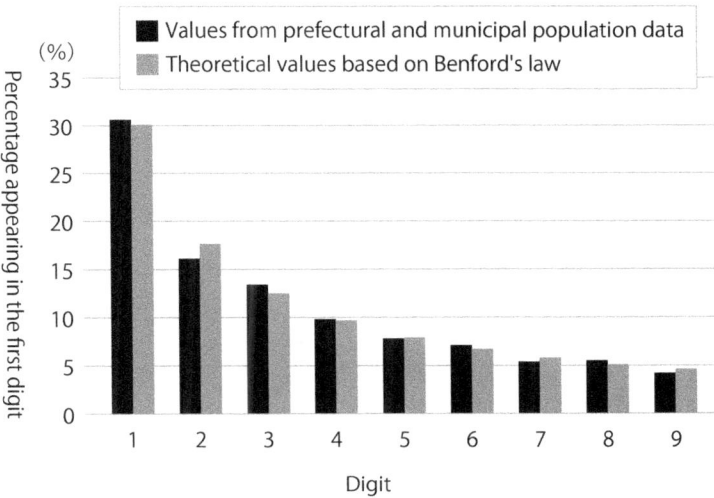

Figure 3.4 Comparison with the theoretical values of Benford's law.

When we compare the two distributions, it's surprising to see how closely the observed results from the populations of prefectures and municipalities match the theoretical values predicted by Benford's Law.

Even numbers that seem to appear randomly can follow a hidden pattern—it's quite fascinating, isn't it?

ABOUT LOGARITHMS

The notation $\log_{10} x$ used in Benford's Law represents the power to which 10 must be raised to obtain x. This is called the *common logarithm*.

For example, when $x = 100$, we have:

$$\log_{10} 100 = 2$$

because $100 = 10^2$. Similarly, when $x = 1000$, then:

$$\log_{10} 1000 = 3$$

You can think of it intuitively as the number of times you need to multiply 10 to reach x. The value of $\log_{10} x$ isn't limited to integers—it can also be a decimal. For instance:

$$\log_{10} 50 \approx 1.699 \ldots$$

If we plot the graph of $y = \log_{10} x$, it looks like the figure shown below.

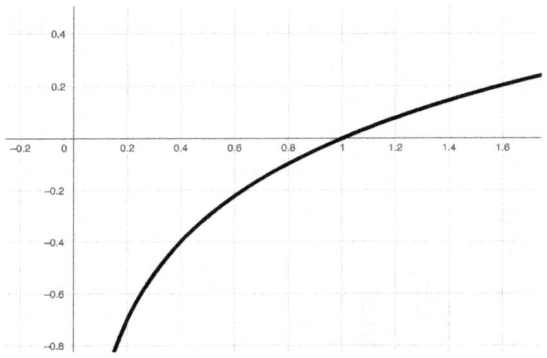

Figure 3.5 The graph of $y = \log_{10} x$.

Since $10^0 = 1$, the graph of $y = \log_{10} x$ passes through the points $(1, 0)$, $(10, 1)$, and $(100, 2)$.

If you zoom out to view a wider range, the curve appears almost flat. This might look quite different from the steep curve you may remember seeing in your high school math textbook.

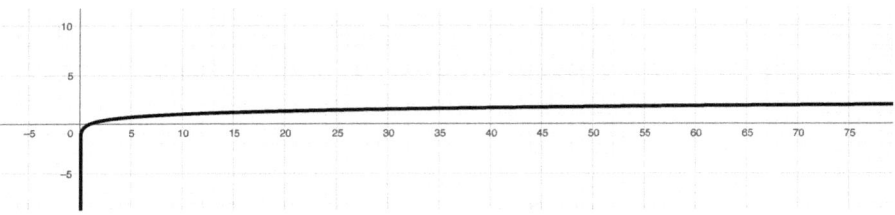

Figure 3.6 Zoomed-out view of the graph of $y = \log_{10} x$.

CHAPTER 4

Conic Sections Created by Light

Figure 4.1 How light from a lamp appears on a wall.

When you shine a flashlight toward a wall in the dark, a bright area appears where the light hits the wall. If you look closely at the boundary between the bright and dark regions, you can find a conic section there.

DOI: 10.1201/9781003670261-4   23

SPREAD OF LIGHT

A typical flashlight has a small bulb inside that emits light. Such a light source is called a *point light source*, and the light spreads out as it travels farther away. In contrast, light that reaches us in parallel rays, like sunlight, is called a *parallel light source*. When viewed from the side, light from a point source looks like a triangle, but the space the light reaches actually forms a cone, as shown in the diagram below.

Figure 4.2 Space reached by light.

The figure shown at the beginning illustrates the shape formed when the cone of light from a point source is blocked by a wall. In other words, it represents the *cross-section* created when a cone is cut by a plane.

CUTTING A CONE

Although cones are simple shapes that we learn about in elementary school, they have fascinating properties. If you cut a cone with a plane and look at the shape of the resulting cross-section, you'll find distinctive curves such as *parabolas*, *hyperbolas*, and *ellipses*. These curves are collectively called *conic sections*.

Because these curves appear as cross-sections of a cone, they are also known as *conic curves*. The relationship between the method of cutting a cone and the types of curves that appear is shown in the following table.

Table 4.1 Relationship Between How a Cone is Cut and the Type of Curve that Appears

How the Cone is Cut	Type of Curve that Appears
Cut by a plane that goes all the way around the side surface	Ellipse (including circles)
Cut by a plane parallel to a generator (slant edge) of the cone	Parabola
All other cases	Hyperbola (If the plane passes through the cone's apex, the result is two straight lines)

The diagram below shows the cone formed by a flashlight beam, with the names of these curves labeled on it.

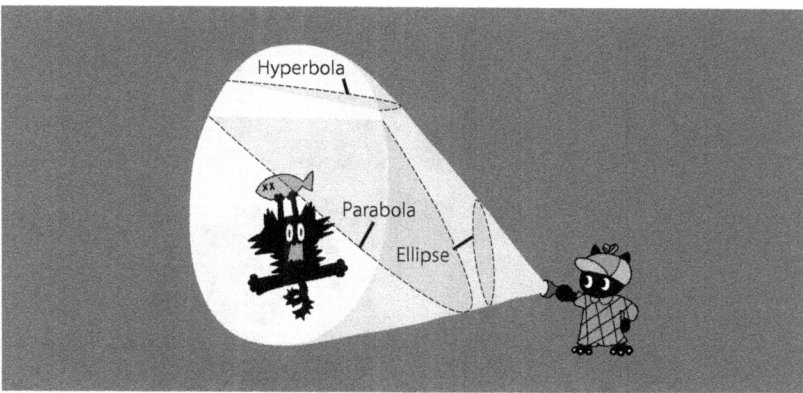

Figure 4.3 Cutting a cone and the curves that appear in its cross-section.

If you take another look at the initial image, you can see that the cone of light created by the flashlight is being cut by a wall. The brightly lit area represents the shape of this cross-section. If the boundary of this cross-section forms a closed loop, the shape will be a circle or an ellipse. On the other hand, if it doesn't form a loop—for example, when the flashlight shines upward from below, as in the opening image—the shape will be a parabola or a hyperbola.

Try shining a flashlight at a wall (or even at the ground) to observe a conic section for yourself. By observing the shapes created by light and shadow in this way, you can discover the fascinating world of mathematics hidden within them.

CHAPTER 5

What Is the True Shape of a Crescent Moon?

Figure 5.1   Crescent moon drawn in different styles.

It might seem a bit nitpicky to criticize whimsical and dreamy illustrations of the moon, but none of the drawings in the image above actually show the correct shape of a crescent moon. Can you point out why this is wrong?

SHAPE OF A CRESCENT MOON

The familiar crescent shape appears when only part of the moon is lit up. The bright area is the portion of the moon's surface that is

                                    DOI: 10.1201/9781003670261-5

illuminated by sunlight. The part that sunlight doesn't reach remains dark and invisible.[1]

Now, let's think about the relationship between the sun and the moon. As shown in the illustration below, the half of the moon facing the sun is brightly lit, while the other half lies in shadow where the sun's light doesn't reach.[2]

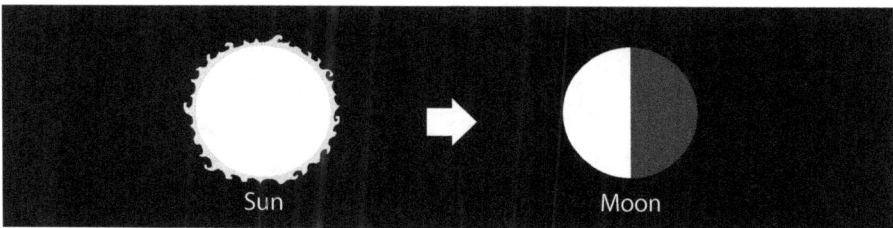

Figure 5.2 Sunlight reaches half of the Moon's surface.

When you observe this situation from different positions, the appearance of the moon changes, as shown below. This is the mechanism behind the moon's phases.

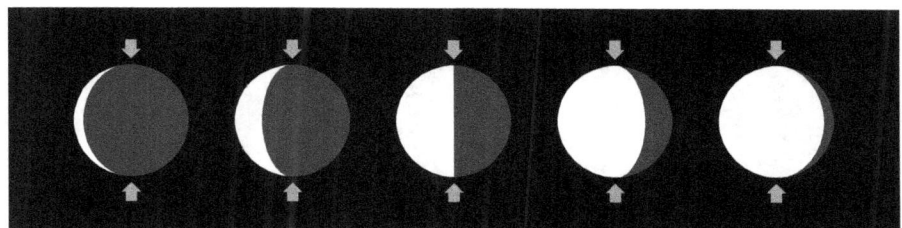

Figure 5.3 Views of the Moon from different angles.

In the figure above, arrows indicate the boundary between the lit and unlit areas along the outline of the moon—that is, along the edge of the circle. If we focus only on the area near the moon's outline, as shown in the diagram below, you'll see that in each case, the bright and dark areas are evenly split. The line connecting the arrows passes through the center of the circle. Also, the boundary of the bright area forms a *great*

---

[1]This time, we'll set aside the special case where the moon enters Earth's shadow—what we call a *lunar eclipse*.

[2]When something blocks light and creates a dark area, that dark part is called a "shadow" On the other hand, a place where light or illumination doesn't directly reach is referred to as "shade"

circle—the largest possible circle that can be drawn on the surface of a sphere.

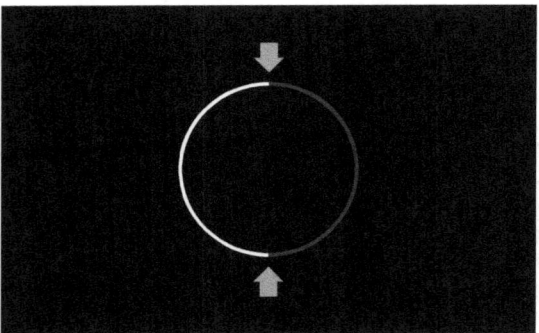

Figure 5.4 Close-up of the outline area.

Even when each of these states is tilted, as shown in the diagram below, the same pattern holds. If you focus on the outline, the bright and dark areas are still evenly divided, and the line connecting the arrows still passes through the center of the circle.

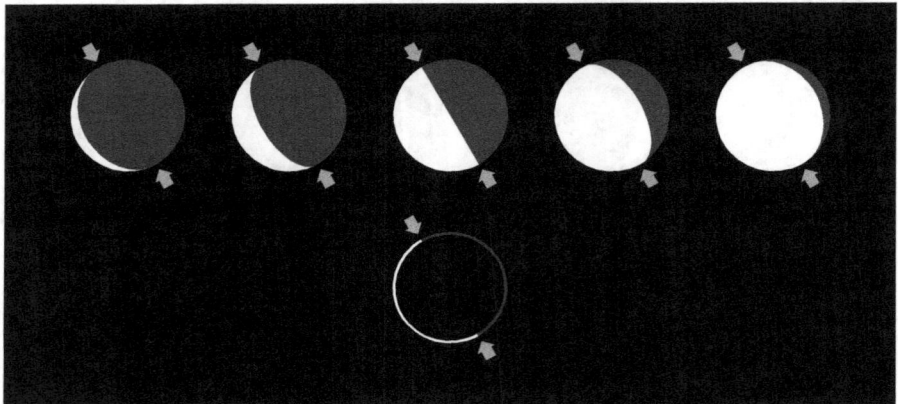

Figure 5.5 Tilted views.

The correct shape of a crescent moon is like the figures shown in the above image. That's why none of the illustrations shown at the beginning can be considered accurate representations of a crescent moon.

That said, a realistic crescent might not look quite as charming. Illustrations often exaggerate features for stylistic effect, so pointing out inaccuracies might seem a bit overly serious. Still, if you understand how light falls on a sphere as described here, you'll be able to draw the correct shape when needed.

By the way, there are even illustrations on the internet—like the one below—showing stars visible in the dark portion of the moon. That's clearly off. The shadowed area of the moon isn't missing; it's just not lit. So you wouldn't see stars through it!

CHAPTER 6

An Equation That Draws the Shape of a Pon de Ring Donut

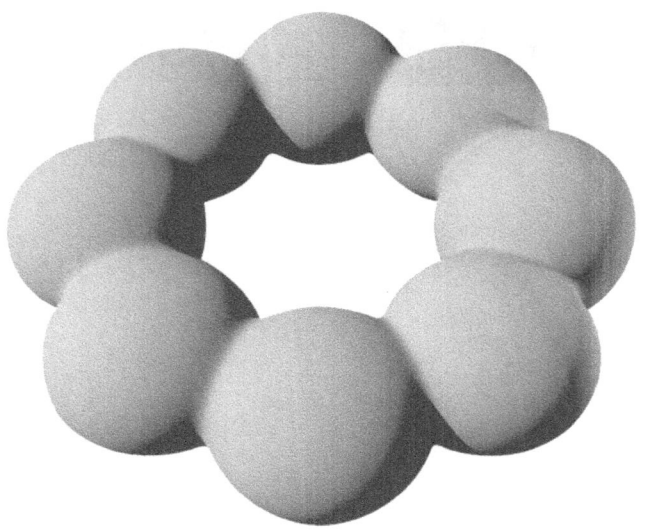

Figure 6.1 A shape drawn by a mathematical equation.

DOI: 10.1201/9781003670261-6

Have you ever seen a donut made up of eight round shapes connected together, known as "Pon de Ring"?[1] The shape shown above, which closely resembles it, was drawn by a computer using the following equation:

$$\sum_{k=0}^{7} \exp\left\{-0.7\left(\left(x - 6.75\cos\frac{k\pi}{4}\right)^2 + \left(y - 6.75\sin\frac{k\pi}{4}\right)^2 + 1.2z^2\right)\right\}$$
$$-0.001 = 0.$$

The summation symbol $\sum_{k=0}^{7}$ means we're summing the values of the expression for $k = 0$ to 7. The function exp() represents the exponential function with base e (*Euler's number*[2]).

If you plot all the combinations of x, y, z that satisfy this equation in three-dimensional space, you'll get a shape that looks just like a donut. Specifically, it resembles the Pon de Ring donut. If we define a function of three variables x, y, z as $F(x, y, z)$, then the equation $F(x, y, z) = 0$ can be used to describe the shape of a solid figure.

When you hear the word "function," many of you might recall drawing various graphs of functions in middle or high school math classes.

Have you ever noticed how the curves of those graphs sometimes looked like actual shapes? By cleverly designing a function with variables x and y, you can create graphs that form illustrations. And by extending this idea to functions of x, y, z, you can create three-dimensional surfaces and shapes.

SHAPES DRAWN BY FUNCTIONS

There's even a form of art often referred to as "function art" (or "mathematical art"). It involves expressing interesting shapes using $y = f(x)$ equations.

For example, the function

$$f(x) = x^2$$

draws a parabola like this:

---

[1] In Japan, it's a popular item sold at a donut chain called Mister Donut.

[2] Euler's number is approximately 2.7182818..., with its decimal part continuing infinitely without repeating. Like the number π, it is known as a *transcendental number*.

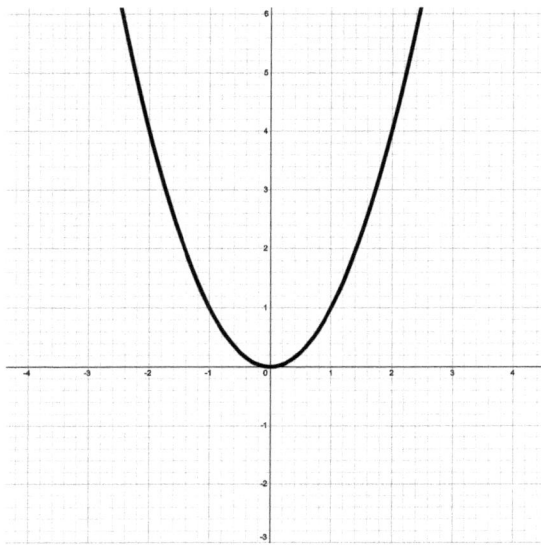

Figure 6.2 Graph of the function $f(x) = x^2$.

The graph of the function

$$f(x) = x^5 - 3x^3 + 1$$

has the following shape.

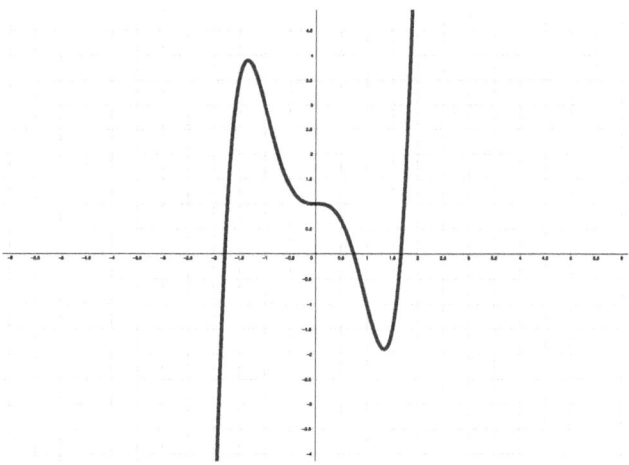

Figure 6.3 Graph of the function $f(x) = x^5 - 3x^3 + 1$.

Since the function can generate various types of graphs, it feels like you could create some interesting forms just by experimenting with different formulas.

If you want to input a formula and instantly see its graph, free software tools like **GeoGebra** (used for geometric and algebraic calculations) or **Desmos** (specialized in equation graphing) are highly recommended. Just type in your formulas, and the graph will appear on the screen.

By the way, did you know that functions can be categorized into *explicit functions* and *implicit functions*?

An explicit function is written in the familiar form

$$f(x) = (\text{an expression in } x),$$

which you may have learned in middle or high school. Once a value for x is given, you can directly compute the corresponding value. The two graphs we saw earlier are both examples of using explicit functions as follows

$$y = f(x),$$

where the shape of the graph is determined by the expression $f(x)$.

However, consider an equation like

$$x^2 + y^2 - 1 = 0.$$

This can be written as

$$(\text{an expression involving } x \text{ and } y) = 0.$$

Equations like this are called *implicit equations*. From such an equation, one can sometimes define y as an *implicit function* of x. In general, an implicit equation is expressed in the form

$$F(x, y) = 0$$

using a function $F(x, y)$ with x and y as variables.

With the form $y = f(x)$, each value of x corresponds to only one value of y. This limitation makes it difficult to represent certain shapes, such as closed curves like circles. However, implicit equations can represent a wide variety of shapes.

For example, the following equation is known as an implicit equation that represents the shape of a heart:

$$(x^2 + y^2 - 1)^3 - x^2 y^3 = 0.$$

When you plot the set of x and y values that satisfy this equation on the xy-plane, a heart-like shape appears as shown below.

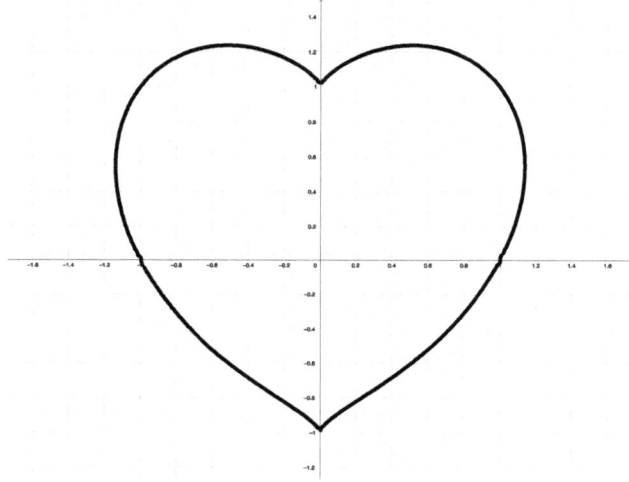

Figure 6.4 Graph of the implicit equation $(x^2 + y^2 - 1)^3 - x^2 y^3 = 0$.

As you can see, using implicit equations allows for the representation of a wider variety of shapes. Now, let's take the left side of the heart equation from earlier and use it as follows:

$$z = (x^2 + y^2 - 1)^3 - x^2 y^3.$$

This equation defines z as a value that depends on x and y, so you can think of it as a graph in three-dimensional space. When you plot this graph, you get a 3D surface like the one shown below.

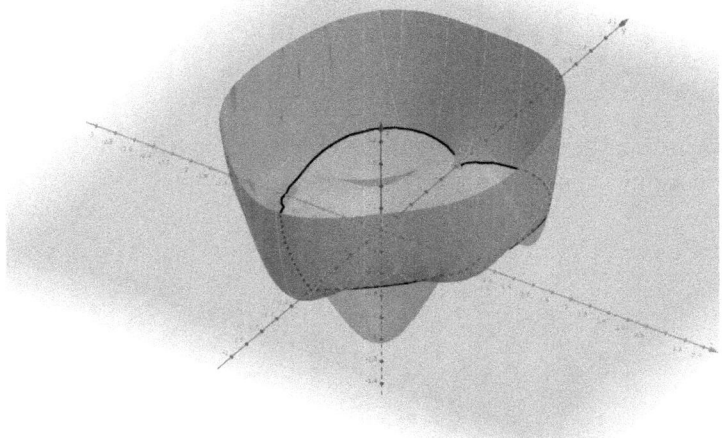

Figure 6.5 Graph of the function $z = (x^2 + y^2 - 1)^3 - x^2 y^3$.

When you slice this 3D shape along the xy-plane, the cross-section you see corresponds to the set of x and y values where $z = 0$. That's exactly where the heart shape appears.

EXTENSION TO THREE DIMENSIONS

With an implicit equation in the form

$$F(x, y) = 0,$$

we were able to draw two-dimensional line shapes. By extending this idea to three dimensions, we can create three-dimensional figures.

Specifically, we define a function $F(x, y, z)$ using the three variables x, y, and z, and plot the set of points where the function equals zero:

$$F(x, y, z) = 0.$$

For example, if we use

$$x^2 + y^2 + z^2 - 1 = 0,$$

this represents a sphere of radius 1 centered at the origin.

Using the equation[3]

$$(x^2 + \tfrac{9}{4}y^2 + z^2 - 1)^3 - x^2z^3 - \tfrac{9}{80}y^2z^3 = 0,$$

you can draw a three-dimensional heart shape like the one shown below. It's quite a cute shape, isn't it?

Figure 6.6 Heart shape represented by an equation with x, y, and z.

Now, let's take another look at the equation that defines the shape of the Pon de Ring we saw at the beginning. The equation itself is quite complex. What follows is an explanation for those who are curious about the details.

The part of the equation that contains exp(expression) is in the form of what's known as a Gaussian function. For example, in physics, this kind of function is sometimes used to describe things like the density of an electromagnetic field in three-dimensional space.

In the earlier equation, eight sources of such a field are arranged evenly along a circular path, and their values are added together. Then, by plotting the locations where that total reaches exactly 0.001, the shape resembling a Pon de Ring is created.

---

[3]Source: Heart Surface - Wolfram MathWorld https://mathworld.wolfram.com/HeartSurface.html

By understanding the properties of various mathematical functions, you can combine that knowledge to enjoy what's called *function art*. Even without diving into the technical details, simply playing around with a graphing app can lead to surprising and fun shapes—it's a great way to explore!

FUNCTION ART

With the rise of software that can instantly draw graphs from mathematical equations, more and more people are enjoying what's known as "function art"—creating illustrations using mathematical functions.

The idea of representing a Pon de Ring using functions was first introduced by CHARTMAN, who is active on X (formerly Twitter), in the following post:

https://x.com/CHARTMANq/status/946000672746487808

CHAPTER 7

The Method is Wrong, but the Result is Correct

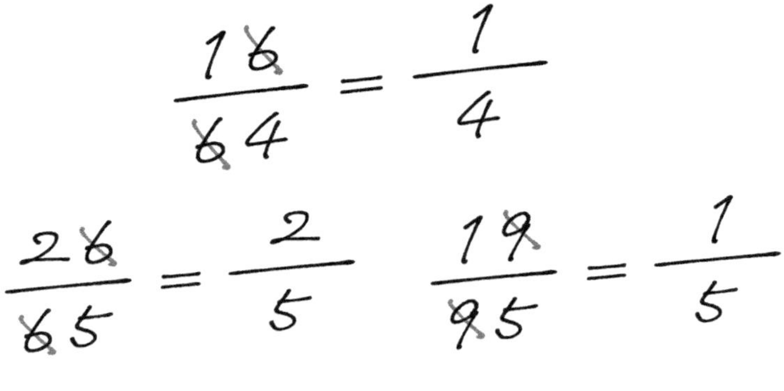

$$\frac{1\cancel{6}}{\cancel{6}4} = \frac{1}{4}$$

$$\frac{2\cancel{6}}{\cancel{6}5} = \frac{2}{5} \qquad \frac{1\cancel{9}}{\cancel{9}5} = \frac{1}{5}$$

Figure 7.1 Simplifying a fraction using the wrong method but still getting the correct result.

$\frac{2}{6}$ is equal to $\frac{1}{3}$, since both the numerator and denominator are divided by the same value, 2. Similarly, $\frac{4}{8}$ is equal to $\frac{1}{2}$, as both the numerator and denominator are divided by 4. When you divide the numerator and denominator by the same number, the value of the fraction remains unchanged. This process of dividing both the numerator and denominator by the same number to make the denominator smaller is called *simplification* (or *reducing a fraction*). Often, simplification refers specifically to making the denominator as small an integer as possible.

By the way, the illustration at the beginning clearly shows an incorrect method of simplification. It compares the numerator and the denominator and simply cancels out matching digits. This is definitely

 DOI: 10.1201/9781003670261-7

not the correct way to simplify a fraction. And yet, somehow, the result ends up being the same as if it had been simplified properly.[1] Why is that?

SIMPLIFYING FRACTIONS

To simplify a fraction, the key is to find a common factor—a number that evenly divides both the numerator and the denominator.

Let's try simplifying the fractions shown in the illustration from the beginning. For $\frac{16}{64}$, if you notice that both the numerator and denominator are divisible by 16, you can simplify it to $\frac{1}{4}$. For $\frac{26}{65}$, since both numbers are divisible by 13, the simplified form is $\frac{2}{5}$. And for $\frac{19}{95}$, since 19 divides both the numerator and denominator evenly, it simplifies to $\frac{1}{5}$.

Once you write it down, it may seem simple, but in reality, it can be difficult to find the common factor, which is why many people find simplification challenging.

AN INCORRECT METHOD OF SIMPLIFICATION

Something strange is going on in the illustration at the beginning. It seems like someone has misunderstood how simplification works and believes that you can just cancel out any digits that appear in both the numerator and the denominator.

$$\frac{1\cancel{6}}{\cancel{6}4} = \frac{1}{4} \qquad \frac{2\cancel{6}}{\cancel{6}5} = \frac{2}{5} \qquad \frac{1\cancel{9}}{\cancel{9}5} = \frac{1}{5}$$

In both the left and center examples, the number 6, which appears in both the numerator and the denominator, is canceled out. In the right example, the number 9 is canceled out in the same way, as it is common to both the numerator and the denominator.

Surprisingly, the result ends up being correct! Who would've thought there could be such a simple way to simplify fractions? Let's check whether the same trick works with other fractions too.

$$\frac{3\cancel{9}}{1\cancel{3}} = \frac{9}{1} \qquad \frac{4\cancel{8}}{2\cancel{4}} = \frac{8}{2}$$

[1]This example of incorrect simplification is introduced in "The Book of Curious and Interesting Puzzles" by David Wells.

Uh-oh, both of these turn out to be incorrect. It seems that simply canceling out matching digits doesn't actually work. So why did the first example happen to work?

The truth is, it's simply a matter of chance—those examples were specially chosen because they just happen to give the correct answer, even when simplified using the wrong method.

For example, there are 8,100 possible combinations of two-digit numerators and denominators (ranging from 10 to 99). But if you search through all of them to find cases where this incorrect digit-canceling method gives the right answer, only the three shown at the beginning turn out to work.

In other words, these are extremely rare, special cases. You might have thought, "Wow, this is a revolutionary and super easy way to simplify fractions!"—but unfortunately, that's not the case.

WHAT ABOUT THREE- OR FOUR-DIGIT NUMBERS?

The earlier examples involved numerators and denominators with two digits. So what happens with three-digit numbers?

Out of curiosity, I wrote a small program and ran it on a computer. It turns out there are four such cases where this mistaken simplification method still gives the correct answer.

$$\frac{1\cancel{66}}{\cancel{66}4} = \frac{1}{4} \qquad \frac{2\cancel{66}}{\cancel{66}5} = \frac{2}{5}$$

$$\frac{4\cancel{84}}{\cancel{84}7} = \frac{4}{7} \qquad \frac{19\cancel{9}}{\cancel{9}95} = \frac{1}{5}$$

Except for the one in the lower left, the others are quite similar to the two-digit cases. I actually expected to find more, but there were only four such examples with three-digit numbers.

Since we've come this far, I went ahead and extended the search to four-digit numbers as well. This time, I found seven such cases as shown below. I used a computer for this search, but trying to find these by hand would definitely be a tough task!

$$\frac{1666}{6664} = \frac{1}{4} \qquad \frac{2666}{6665} = \frac{2}{5} \qquad \frac{1999}{9995} = \frac{1}{5}$$

$$\frac{1957}{2575} = \frac{19}{25} \qquad \frac{2369}{6695} = \frac{23}{65}$$

$$\frac{1378}{6784} = \frac{13}{64} \qquad \frac{2678}{9785} = \frac{26}{95}$$

A MISUNDERSTANDING IN LOGARITHMIC CALCULATIONS

A similar kind of misunderstanding can happen with *logarithmic calculations*. Take a look at the following two equations—they're both correct:

$$\log 1 + \log 2 + \log 3 = \log 6$$

$$\log 2 + \log 2 = \log 4.$$

In these cases, we're omitting the base of the logarithm, since it can be any value. Looking at these examples, you might mistakenly think, "Oh, I just need to add the numbers inside the logs." As a result, you might be tempted to write:

$$\log 1 + \log 3 = \log 4.$$

But that's incorrect. The correct calculation is

$$\log 1 + \log 3 = \log 3.$$

If we express it with variables a and b, we have:

$$\log a + \log b = \log(ab).$$

So, instead of adding the numbers inside the logs, you should actually multiply them. Just like in the case of simplifying fractions, if you don't fully understand the correct method, you might see an exceptional case and end up learning the wrong rule from it.

SLIDE RULE

If we are in a situation where it's easy to switch between a number x and its logarithm $\log x$, we can use the logarithmic identity

$$\log a + \log b = \log(ab)$$

to turn multiplication into addition.

For example, if we want to find the product c of two numbers a and b, instead of multiplying them directly, we can first take the logarithms of a and b, then add them:

$$\log a + \log b.$$

Since this sum equals $\log c$, we can then reverse the logarithmic operation—that is, apply the *exponential function*—to get back the original number c.

In short, multiplying two numbers can be turned into an addition problem by converting the numbers into their logarithms and back.

Before calculators existed, multiplying large numbers was a time-consuming and challenging task. To solve this, a tool called the "slide rule" was developed, which applied the idea of using logarithms to transform multiplication into addition. By sliding bars and reading the scales, the slide rule allowed for intuitive and efficient conversion between numbers and their logarithms. We won't go into the details of how slide rules work here, but if you're curious, it's definitely worth looking into. While they've fallen out of use due to the spread of calculators, slide rules were once essential tools for engineers.

CHAPTER 8

A Mathematical Mystery! The Collatz Conjecture

Pick any integer greater than 1, and try repeating the following steps:

- If the number is even, divide it by 2.

- If the number is odd, multiply it by 3 and add 1.

If you keep doing this, it seems that no matter what number you start with, you'll always end up reaching 1.

But is that really true? So far, no one has been able to prove it. At the same time, no one has found a counterexample either. This puzzle is known as the *Collatz Conjecture*, named after the German mathematician Lothar Collatz. It's a famously difficult mathematical problem—so much so that proving it could earn you a reward of about 830,000 dollars!

DOI: 10.1201/9781003670261-8

THE COLLATZ CONJECTURE

The rules, as mentioned earlier, are very simple. Let's pick a number and give it a try.

Suppose we start with 9. Since 9 is odd, we multiply it by 3 and add 1, which gives us 28. Next, 28 is even, so we divide it by 2 to get 14. Since 14 is also even, we divide it by 2 again to get 7. Now 7 is odd, so we multiply by 3 and add 1 to get 22, and so on.

Following this pattern, we can observe the sequence progressing like this:

$9 \to 28 \to 14 \to 7 \to 22 \to 11 \to 34 \to 17 \to 52 \to 26 \to 13 \to 40 \to 20 \to 10 \to 5 \to 16 \to 8 \to 4 \to 2 \to 1.$

It took 19 steps in total, so it turned out to be more work than expected. The values don't just decrease steadily—they go up and down before finally reaching 1.

Now, what happens if we try 15? Following the same rules, we eventually encounter the number 40 in the process:

$15 \to 46 \to 23 \to 70 \to 35 \to 106 \to 53 \to 160 \to 80 \to 40.$

Since 40 already appeared in the sequence we saw earlier when we started with 9, we can be sure that this sequence will also eventually reach 1.

In this way, we can see that the Collatz Conjecture holds true for the numbers 9 and 15. But what about other numbers? Give it a try yourself!

If you happen to choose 27, you're in for a long ride. In that case, it takes 111 steps to complete the process. At one point, the number even jumps up to 9,232 before eventually reaching 1.

The Collatz Conjecture remains an unsolved problem in mathematics—despite its incredibly simple rules, no one has yet been able to prove it. That might come as a surprise. Although it hasn't been proven for every possible number, extensive testing has confirmed that the rule holds true for all numbers up to 2^{68}, which is 295,147,905,179,352,825,856.

Still, we can't rule out the possibility that a number larger than this might behave differently and never reach 1. So, if you're looking to find a counterexample, try testing numbers even larger than that!

It's said that if you can prove the conjecture—or find a counterexample—you could earn a reward of 120 million yen.[1] Isn't it exciting that something so simple has stumped so many mathematicians for so long?

---

[1] In 2021, a Japanese company announced a reward of 120 million yen (about 830,000 dollars) for solving the unsolved math problem known as the Collatz Conjecture.

Strange Magic Square

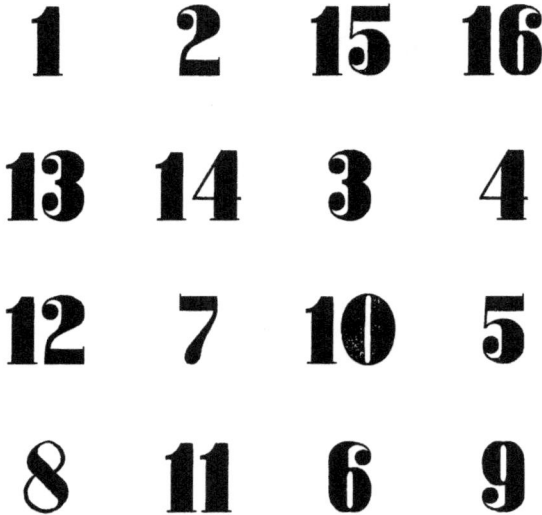

Figure 9.1 The 4×4 magic square.

A *magic square* is a grid of $n \times n$ cells in which numbers are arranged so that the sum of the numbers in each row, column, and diagonal is the same.

Typically, as shown in the illustration above, a 4×4 magic square uses the numbers from 1 to 16, each exactly once. The sum of the numbers in any row, column, or diagonal is always 34. For example, if we add the numbers in the first row: $1 + 2 + 15 + 16 = 34$. Likewise, for the diagonal from the top-left to the bottom-right: $1 + 14 + 10 + 9 = 34$. Try checking other rows and columns too!

DOI: 10.1201/9781003670261-9

THE MAGIC SQUARE AT THE SAGRADA FAMÍLIA

The image below illustrates a magic square found on the wall of the famous architectural masterpiece, the Sagrada Família, designed by Gaudí. This square has some unusual and intriguing features that set it apart from typical magic squares.

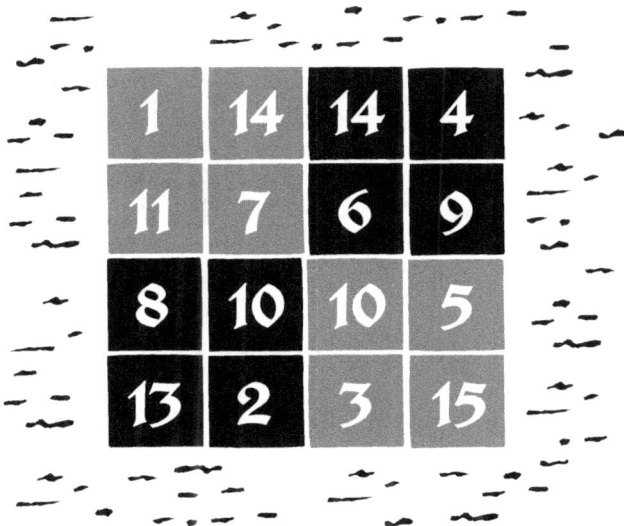

Figure 9.2 The stone magic square at the Sagrada Família.

If you look closely, you'll notice that the numbers 12 and 16 are missing. Instead, the numbers 10 and 14 appear twice. Nevertheless, the sum of the numbers in each row, column, and diagonal remains the same. As you can see by calculating it, the sum is 33 (interestingly, that's the number of topics covered in this book!).

But that's not all—each of the 2 × 2 sections you get by dividing the square into four parts also adds up to 33. And even more impressively, if you shift this 2 × 2 section one square to the right or down, wrapping around to the opposite side as shown in the next diagram, the sum still remains 33 in every case. It's an incredibly well-crafted design!

WAYS TO MAKE 33

Are there other patterns where the sum of four numbers adds up to 33?

Figure 9.3 Ways to group four cells that sum to 33.

A computer search revealed that there are 389 different ways to divide the square into groups of four cells such that the sum in each group is 33. At this point, we're moving quite far from the original idea of a magic square that focuses on rows, columns, and diagonals. Even so, some of these groupings are quite interesting, like the one shown in the figure below.

Figure 9.4 Unique groupings of four cells that sum to 33.

In each pattern, cells of the same color form a group of four, and the sum of the numbers in each group is 33. Try checking them yourself!

Was this level of design actually intended, or is it just a coincidence? The mystery only deepens.

HOW TO CREATE A STANDARD 4×4 MAGIC SQUARE

Unlike the unique magic square found on the Sagrada Família's stone slab, a standard 4×4 magic square uses the numbers from 1 to 16, each exactly once. It's known that there are a total of 880 such magic squares.

While it's possible to create one by trial and error—rearranging the numbers from 1 to 16 until everything lines up—there's actually a simple method for constructing one, as introduced below.

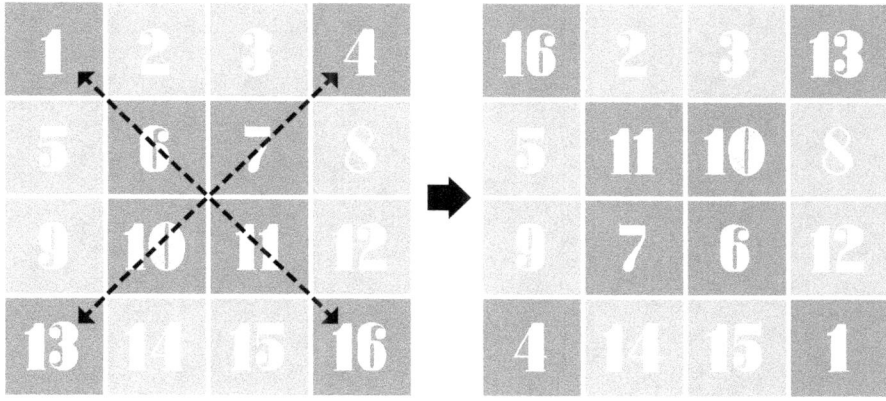

Figure 9.5 A simple way to create a 4×4 magic square.

First, as shown in the left diagram above, begin by filling in the numbers from 1 to 16 in order, starting from the top-left cell and moving to the right. When you reach the end of a row, continue from the leftmost cell of the next row.

Next, as illustrated on the right, look at the two diagonals and reverse the order of the numbers on each. For the diagonal from the top-left to the bottom-right, which originally had the numbers $1, 6, 11, 16$, reverse them to get $16, 11, 6, 1$. Do the same for the diagonal from the top-right to the bottom-left: reverse $13, 10, 7, 4$ to get $4, 7, 10, 13$.

Now, with these new arrangements, does the grid form a magic square? What about if you divide it into four 2×2 sections—do those sections also have equal sums? Be sure to check!

By the way, for a 5 × 5 magic square using the numbers 1 through 25, it is known that there are exactly 275,305,224 different possible configurations.

CHAPTER 10

Rings That Look Different but Have the Same Volume

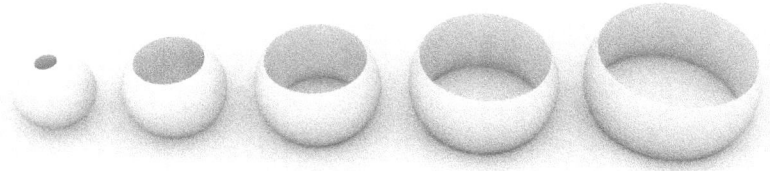

Figure 10.1 Five rings of different sizes but equal volume.

The five rings shown in the image above all differ in size. However, the width of each ring is the same. So, which ring has the largest volume? At first glance, the one on the far right might seem the biggest. Surprisingly, though, they all have the same volume.

THE NAPKIN RING PROBLEM

These five rings were created as follows: First, five spheres of different sizes were prepared.

DOI: 10.1201/9781003670261-10

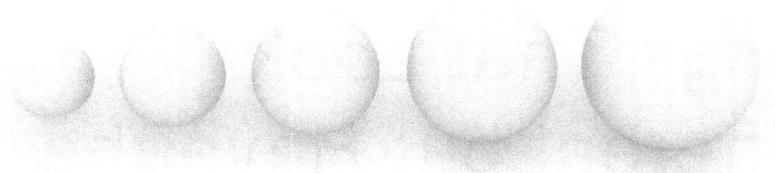

Figure 10.2 Five spheres of different sizes.

Each sphere is hollowed out using a cylindrical drill to create a ring shape. The thickness of the cylinder is adjusted so that the resulting rings all have the same width.

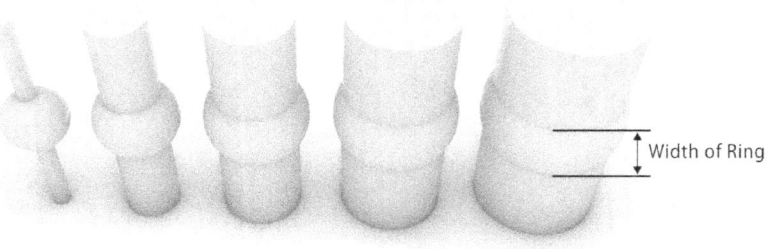

Width of Ring

Figure 10.3 Each sphere is hollowed out with a cylinder of different thickness.

Even though the resulting rings look quite different in size, they all end up having the same volume—how curious! While we've been calling these shapes "rings," they closely resemble napkin holders, which is why this is known as the "Napkin Ring Problem."

If we let the width of the ring be w, and the original sphere's radius be r, then the volume of the ring turns out to be:

$$\frac{\pi w^3}{3}$$

For the derivation of the formula, please refer to the explanation of the Napkin ring problem on Wikipedia.[1] Interestingly, this formula does not include r, the radius of the original sphere. This means that the volume of the ring depends only on the width w. Since all the rings have the same width, they must also have the same volume.

Now, let's imagine what happens if the cylinder's thickness becomes extremely small. In that case, the ring shape approaches that of a complete sphere, and the ring's width becomes equal to the sphere's diameter. The volume of such a ring can then be considered as the volume of a sphere with radius $\frac{1}{2}w$. Substituting this into the formula for the volume of a sphere, $\frac{4}{3}\pi r^3$, we get:

$$\frac{4}{3}\pi\left(\frac{1}{2}w\right)^3 = \frac{\pi w^3}{6}.$$

This is indeed equal to the previous result. Everything checks out!

THE LENGTH OF A STRING RAISED ABOVE THE SURFACE OF A SPHERE

Let's look at another counterintuitive problem, similar in spirit to the Napkin Ring Problem.

Imagine you have a large globe with a diameter of 10 meters. You place a string snugly around its equator. The length of this string is the circumference of the globe: $\pi \times 10 \approx 31.4$ meters. Now, suppose you lift this string 50 cm (0.5 meters) above the surface all the way around.

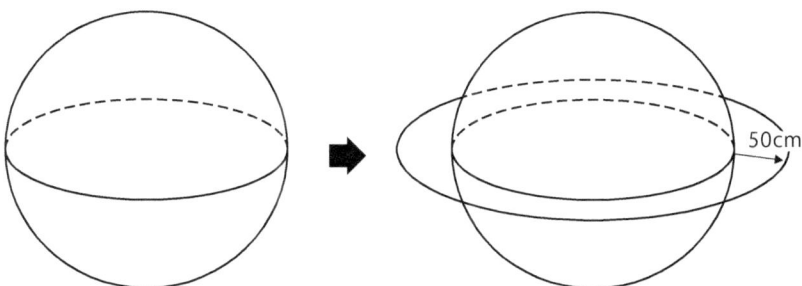

Figure 10.4   Lifting the string 50 cm above the equator.

The new string will now form a circle with a diameter of 11 meters.

[1] https://en.wikipedia.org/wiki/Napkin_ring_problem

Its circumference becomes $\pi \times 11 \approx 34.54$ meters. So you would need an additional:

$$34.54 - 31.4 = 3.14 \text{ meters.}$$

Now, let's do the same for the actual Earth, which has a circumference of about 40,000 km. You place a string around the equator and then raise it 50 cm above the surface. How much extra string do you need?

Let the Earth's radius be r meters. The original length of the string is $2\pi r$. After raising the string 0.5 meters, its new length is $2\pi(r + 0.5)$. The increase in length is

$$2\pi(r + 0.5) - 2\pi r = \pi \text{ meters.}$$

So regardless of the size of the sphere—even if it's Earth itself—the string needs to be about 3.14 meters longer.

Surprising, isn't it? Our intuition often leads us astray!

CHAPTER 11

Convex Polygons and Radar Charts

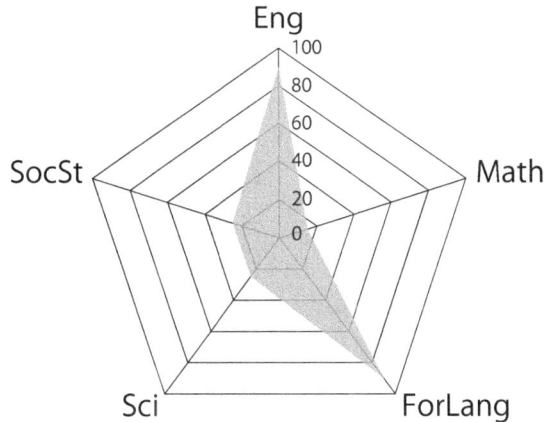

Figure 11.1 Example of a radar chart.

The graph shown above, which often appears in score reports from mock exams or sports tests, is called a *radar chart*. It visually highlights which subjects a student excels in and which ones need improvement. The shape of the figure also gives an intuitive sense of how balanced the overall performance is, and the area can suggest the total score at a glance. It seems like a great way to present information—but is it really all good?

DOI: 10.1201/9781003670261-11

THE INFLUENCE OF SUBJECT ORDER

Following radar charts are for a student's scores in five subjects: English 90, Mathematics 15, Science 90, Social Studies 25, and Foreign Language 25. Both charts use the same data; the only difference is the order in which the subjects are arranged.

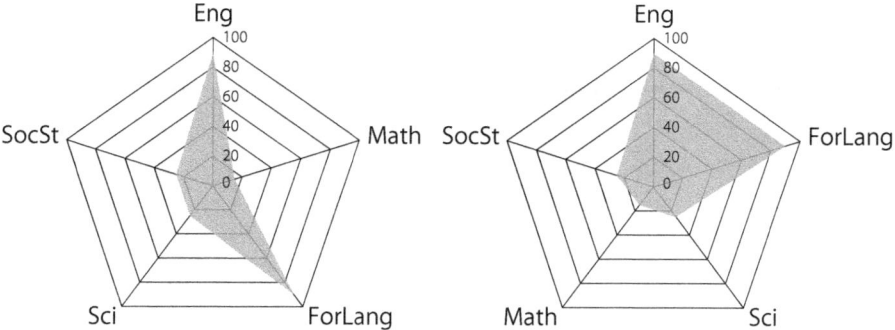

Figure 11.2 Two radar charts with different subject orders.

Compared to the left chart, the right one appears to have no dips and looks more balanced overall. But what about the area?

The area of a five-subject radar chart can be calculated by dividing it into five triangles, each with the center of the graph (where the score is 0) as one of its vertices. Using the formula for the area s of a triangle with two sides a, b, and an angle θ between them:

$$s = \frac{1}{2}ab\sin\theta$$

the total area S of the radar chart can be calculated as:

$$S = \frac{1}{2}(ab + bc + cd + de + ea)\sin 72°.$$

Here, a through e represent the scores for each subject.

In other words, the total area depends on the sum of the products of adjacent subject scores. Using this calculation, it turns out that the right-hand chart has an area 1.7 times larger than the left-hand one.

This means that the way a radar chart looks can change depending on the order of the subjects, and that change can alter the impression it gives. So, it's important to use radar charts with care.

RADAR CHARTS AND CIRCULAR PERMUTATIONS

When there are n items, how many ways can you arrange them on a radar chart? Arranging things around a circle is called a *circular permutation*, and the number of such arrangements is given by $(n-1)!$.

For example, with 3 items there are 2 ways, with 4 items there are 6 ways, and with 5 items there are 24 ways to arrange them.

By the way, when there are 3 items, the radar chart forms a triangle. Although there are 2 ways to arrange the items, one triangle is just a flipped version of the other, so the area remains the same either way.

However, when there are 4 or more items, the area can change depending on the order. That means you can potentially present your results more favorably by choosing the order that gives the largest area. For instance, if there are 10 items, the number of circular permutations is a whopping 3,628,800. Checking all of them would be overwhelming. Problems like this, where the number of combinations grows explosively as the number of items increases, are often referred to as cases of *combinatorial explosion*.

CHAPTER 12

What If We Used Dice to Decide How Much New Year's Money to Give?

	Second Time (y)					
	1	2	3	4	5	6
1	1	1	1	1	1	1
2	2	4	8	16	32	64
3	3	9	27	81	243	729
4	4	16	64	256	1,024	4,096
5	5	24	125	625	3,125	15,625
6	6	36	216	1,296	7,776	46,656

(Row labels under "First Time (x)": 1, 2, 3, 4, 5, 6)

Figure 12.1 The value of x^y when the first roll is x and the second roll is y.

In Japan, there's a tradition called *otoshidama*, where children receive money as a gift from their parents and relatives during the New Year holidays. For parents, however, it can be tricky to decide just how much to give. What if we left it up to chance?

DOI: 10.1201/9781003670261-12

Suppose we let the child roll a die, and the number that comes up determines the amount they receive. It might feel a bit like gambling and not entirely realistic—but as a thought experiment, it's quite a fun idea!

Let's dive right in and think about how we might match the result of rolling dice with the amount of money given. What kind of rules would make this interesting?

The table above shows one idea: you roll two six-sided dice. Let the result of the first roll be x, and the second be y. Then, you receive x^y yen—that is, x raised to the power of y.

Surprisingly, this creates a rather unexpected distribution of possible amounts!

DECIDING NEW YEAR'S MONEY WITH DICE

If we were to decide the amount of New Year's money by rolling dice, it would certainly catch a child's attention—after all, their money is on the line! It might even inspire them to take *probability* and *expected value* more seriously.

Let's start with a simple rule: roll one die, and multiply the number by 1,000 yen. So, if the die shows 1, you get 1,000 yen; if it shows 6, you get 6,000 yen. In this case, what amount can a child expect to receive on average?

To find the expected value, we calculate the sum of (probability) × (amount). Since each number from 1 to 6 has an equal chance of appearing (1/6), we get:

$$\frac{1}{6} \times 1000 + \frac{1}{6} \times 2000 + \cdots + \frac{1}{6} \times 6000 = 3500.$$

So, the expected value is 3,500 yen.

Now, what if we roll two dice and take the sum of the two results, then multiply that by 1,000 yen? This is essentially the same as rolling one die twice and adding the amounts—so the expected value becomes:

$$3500 \times 2 = 7000 \text{ yen.}$$

But since we're already rolling two dice, why not try something a bit more imaginative? This time, let's say the first roll gives a value x

and the second y, and the child receives x^y yen—that is, x raised to the power of y.

For example, if the first roll is 2 and the second is 3, the result is

$$2^3 = 8 \text{ yen.}$$

That's quite a small amount. But what if both rolls show 6? Then the amount is

$$6^6 = 46656 \text{ yen.}$$

Now that's a big payoff!

Looking again at the table showing the possible amounts, we see that if the first roll is between 1 and 3, the maximum is only 729 yen—not very hopeful for the child. But if a 5 or 6 comes up, there's real potential for a jackpot. A 6 followed by another 6 gives 46,656 yen, while a 5 followed by a 6 yields 15,625 yen.

These large numbers stand out, but here's the catch: the expected value of this rule is only 2,283 yen. Despite using two dice, the average amount is lower than the single-die rule of (roll) × 1,000 yen.

So, from the parent's perspective, proposing this "x^y yen" rule might be a clever move. Of course, you might end up having to pay out 46,656 yen—but the odds are in your favor!

STANDARD DEVIATION OF THE AMOUNT

Compared to simply adding the two dice together, using the rule x^y (the first number raised to the power of the second) results in a much greater spread between small and large outcomes. It feels far more like gambling. But how can we actually measure how "risky" or "unpredictable" this method is?

One useful way to measure this is by looking at the *standard deviation*, rather than just the average. Standard deviation tells us how far outcomes tend to deviate from the mean. The formula is

$$\sigma = \sqrt{\frac{1}{n} \sum_{i=1}^{n} (x_i - \mu)^2}$$

Here, n is the number of values, x_i represents each value, and μ is the average (mean).

In plain terms, this is

"The square root of (the average of (the squares of (the differences from the mean)))."

Although we added extra parentheses here to make the relationships clearer, it may be easier to understand by looking at the formula rather than reading the explanation.

Now, let's apply this to our dice experiments: If we take the sum of two dice and multiply it by 1,000, the standard deviation is approximately 2, 415. This means that the amount of money you might receive tends to vary by around ±2400 yen from the average.

If we take x^y, the standard deviation jumps to about 8, 034. (For just one die, multiplied by 1,000, the standard deviation is about 1, 707.)

As you can see, the x^y method introduces a lot more unpredictability—and standard deviation makes that difference easy to grasp.

AND ONE MORE TWIST...

You might be thinking: "Okay, that's enough dice talk." But there's still more fun to be had! This kind of deep thinking about a single idea can lead to surprising insights—and it's fun, too.

For example: What if, after rolling the first die, you were given a chance to switch the formula? Instead of calculating x^y, you could choose to use y^x instead.

The table below shows how much better or worse off you'd be by switching, depending on the first number rolled. A positive value means switching was the right call; a negative one means you should've stuck with the original formula.

	Second Time (y)					
	1	2	3	4	5	6
1	0	1	2	3	4	5
2	-1	0	1	0	-7	-28
3	-2	-1	0	-17	-118	-513
4	-3	0	17	0	-399	-2,800
5	-4	7	118	399	0	-7,849
6	-5	28	513	2,800	7,849	0

(Left axis: First Time (x))

Figure 12.2 Increase or decrease when changing the calculation method for the amount received.

From the table, we can see that if your first roll is a 1 or a 6, it's generally better to switch to y^x. In all other cases, it's best to leave it as x^y. Making this smart decision boosts the expected value of your New Year's money by 933 yen!

This may be getting a bit complex, but when it comes to *otoshidama*, even kids might get seriously interested. Why not use this opportunity to explore probability, averages, expected values—and variance—together?

THE VALUE OF $Y^X - X^Y$

The second table shows the calculated values of $y^x - x^y$. Knowing which combinations of x and y yield a positive result is essentially the same as identifying the region where a graph of

$$z = y^x - x^y$$

lies above the plane $z = 0$—that is, above the xy-plane.

When we actually plot the graph of $z = y^x - x^y$, we see a surprisingly complex shape. This tells us that deciding which is greater—x^y or y^x—isn't such a simple problem after all.

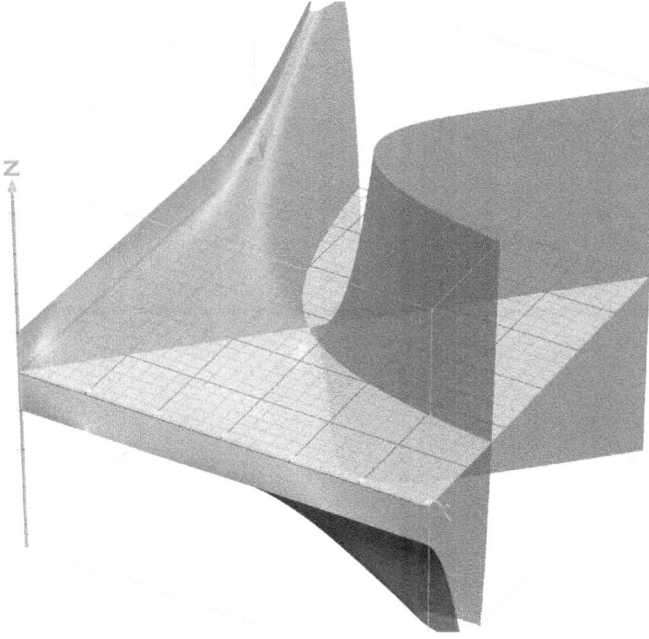

Figure 12.3　Graph of $z = y^x - x^y$.

CHAPTER 13

Shapes Made with Triangles

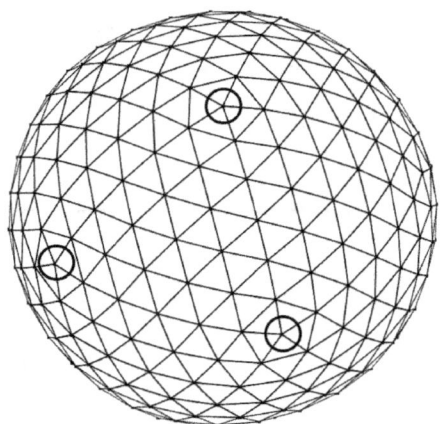

Figure 13.1   A spherical polyhedron composed of triangles.

By combining triangles, we can create various three-dimensional shapes. This approach is especially common in computer graphics (CG), where shapes are often represented as collections of triangles—a method known as *triangle mesh representation*.

Among polygons, triangles are the simplest, making them easy for computers to handle. With just three vertices, a triangle defines a unique plane. In three-dimensional space, any three points will always lie on a flat surface when connected, which is a useful property. (In contrast,

   DOI: 10.1201/9781003670261-13

a quadrilateral formed by connecting four points doesn't necessarily lie flat.) With fine enough triangles, we can flexibly represent a wide variety of shapes.

Now, in the illustration at the beginning, a spherical shape is created entirely out of triangles. At first glance, it appears that nearly identical triangles are neatly tiled across the surface. However, if you look closely, you'll see that the tiling isn't perfectly uniform.

At the vertices circled in the diagram, five triangles meet. At the other vertices, six triangles come together. This raises an interesting question: Is it possible to tile a sphere using triangles such that every vertex is shared by exactly six triangles?

VALENCY OF A POLYHEDRON

A polyhedron consists of vertices, edges, and faces. The number of edges connected to a single vertex is called its *valency*. In the diagram below, the vertex on the left has a valency of 5, while the one on the right has a valency of 6. The number of triangles that share a vertex matches its valency.

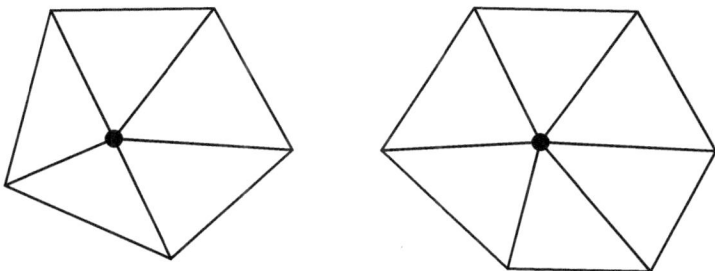

Figure 13.2 Vertexes with valences of 5 and 6.

Using the concept of valency, we can rephrase our earlier question as follows:

"Is it possible to represent a sphere with a triangle mesh in such a way that every vertex has a valency of 6?"

To answer this question, we can turn to a famous theorem about polyhedra.

EULER'S POLYHEDRON THEOREM

It's known that the number of vertices, edges, and faces of a polyhedron always satisfy the following relationship:

$$(\text{Number of vertices}) - (\text{Number of edges}) + (\text{Number of faces}) = 2.$$

This is called *Euler's Polyhedron Theorem*. It holds not only for polyhedra made entirely of triangles but also for those that include quadrilaterals, pentagons, or other types of polygons.

The table below summarizes the number of vertices, edges, and faces for each of the *regular polyhedra* (There are exactly five regular polyhedra: the *regular tetrahedron, cube, regular octahedron, regular dodecahedron,* and *regular icosahedron*). Let's go through them one by one to verify this relationship. You can find that the formula above holds true for all of these.

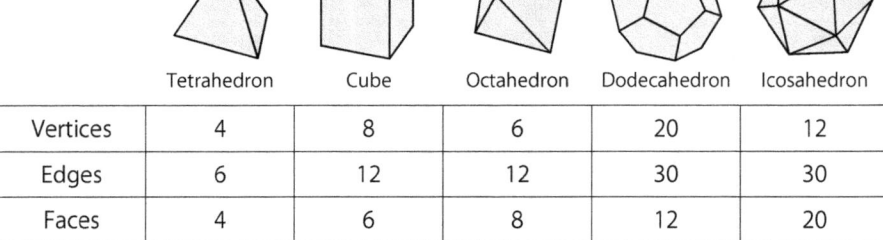

	Tetrahedron	Cube	Octahedron	Dodecahedron	Icosahedron
Vertices	4	8	6	20	12
Edges	6	12	12	30	30
Faces	4	6	8	12	20

Figure 13.3 Number of vertices, edges, and faces of regular polyhedra.

Euler's Polyhedron Theorem doesn't apply only to regular polyhedra. It also holds for more complex shapes like the ones shown in the figures below. Let's plug the number of faces, edges, and vertices into the formula and see.

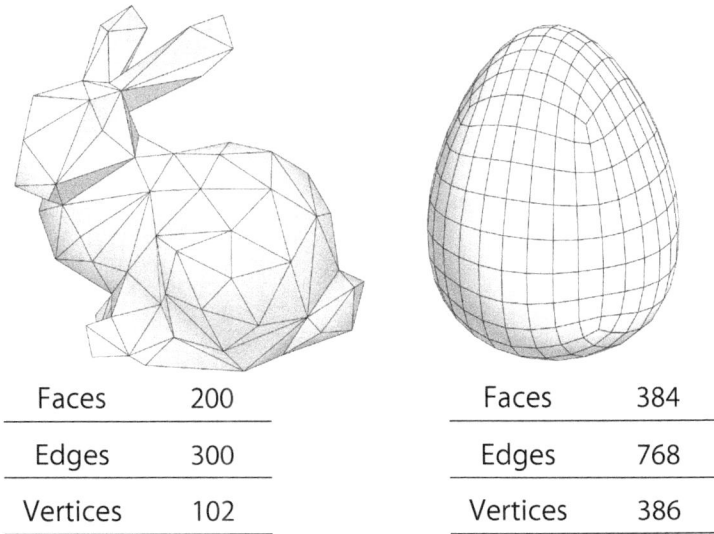

Faces	200
Edges	300
Vertices	102

Faces	384
Edges	768
Vertices	386

Figure 13.4 Euler's polyhedron formula also holds for these solids.

For the bunny-shaped figure on the left:

$$102 \text{ (vertices)} - 300 \text{ (edges)} + 200 \text{ (faces)} = 2.$$

For the egg-shaped figure on the right:

$$386 \text{ (vertices)} - 768 \text{ (edges)} + 384 \text{ (faces)} = 2.$$

It's fascinating, isn't it?

However, it's important to note that Euler's Polyhedron Theorem only holds for shapes that have no holes. For example, if a shape has one hole—like a donut or an inner tube—the right-hand side of the equation becomes 0. If a shape has two holes—like a double-seater swim ring—the right-hand side becomes −2.

Taking the number of holes into account, Euler's formula can be rewritten as:

$$\text{(Number of vertices)} - \text{(Number of edges)} + \text{(Number of faces)}$$
$$= 2 \times (1 - \text{Number of holes}).$$

SPHERICAL POLYHEDRA

Now, let's return to the question we posed at the beginning: "Is it possible to represent a sphere with a triangle mesh such that every vertex has a valency of 6?"

Suppose, for the sake of argument, that this *is* possible. If the polyhedron is made up of n triangles, then the number of vertices would be $\frac{n}{2}$. This is because each triangle has 3 vertices, and each vertex is shared by 6 triangles, so:

$$\text{Number of vertices} = \frac{3n}{6} = \frac{n}{2}.$$

Similarly, the number of edges would be $\frac{3n}{2}$, since each triangle has 3 edges, but each edge is shared by 2 triangles:

$$\text{Number of edges} = \frac{3n}{2}.$$

Now, let's plug these values into Euler's formula:

$$(\text{Vertices}) - (\text{Edges}) + (\text{Faces}) = \frac{n}{2} - \frac{3n}{2} + n = 0.$$

No matter how many triangles we use, the left-hand side always equals 0. But this contradicts Euler's Polyhedron Theorem, which states that the left-hand side must equal 2 (for a sphere with no holes). Therefore, our assumption was incorrect.

The conclusion is: "It is not possible for all the vertices to have a valency of 6 in a triangle mesh representation of a sphere."

This is why, even when we try to neatly divide the surface of a sphere into triangles, we inevitably end up with an uneven distribution. So next time you see a dome-shaped structure made of triangles, take a closer look at the valency of its vertices—you might notice this mathematical quirk in action.

Furthermore, we can also explore a similar kind of question: "Is it possible to represent a sphere using only quadrilaterals in such a way that every vertex has a valency of 4?"

In the diagram below, the sphere is made up of quadrilaterals. However, at the locations marked with circles, we see vertices with a valency of 3. Ideally, we'd like every vertex to have a valency of 4, just like the others. But is this actually possible?

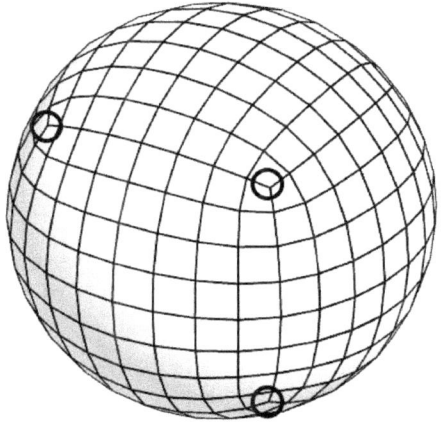

Figure 13.5 A sphere made up of quadrilaterals.

If you've followed along so far, you probably already know how to approach this. Let's assume, as before, that it *is* possible.

Suppose the total number of quadrilaterals is n. Since each quadrilateral has 4 vertices and each vertex is shared by 4 quadrilaterals, the number of vertices would be:

$$\text{Number of vertices} = \frac{4n}{4} = n.$$

Similarly, since each quadrilateral has 4 edges and each edge is shared by 2 quadrilaterals, the number of edges is

$$\text{Number of edges} = \frac{4n}{2} = 2n.$$

Now plug these into Euler's formula:

$$n - 2n + n = 0.$$

No matter how many quadrilaterals we use, the value becomes 0, which contradicts Euler's Polyhedron Theorem. The left-hand side should be 2 for a sphere with no holes. Therefore, our assumption was incorrect.

The conclusion is: "It is not possible for all vertices to have a valency of 4 in a quadrilateral mesh representation of a sphere."

A good real-world example of this is *Geo-Cosmos*, a large spherical display at the Miraikan (National Museum of Emerging Science and Innovation) in Tokyo. Its surface is covered with quadrilateral panels. If you look closely, you'll notice that at certain points, only three panels meet. This is yet another reminder that it's simply impossible to tile a sphere with panels in a perfectly uniform way.

WHAT IF IT'S SHAPED LIKE A DONUT?

For a shape like a donut—with a single hole—Euler's formula becomes:

$$(\text{Number of vertices}) - (\text{Number of edges}) + (\text{Number of faces}) = 0.$$

Interestingly, this is exactly the result we obtained earlier when we *assumed* that a shape was made entirely of triangles with all vertices having a valency of 6. We also got the same result when we assumed a shape was made entirely of quadrilaterals with all vertices having a valency of 4.

In fact, the polyhedra shown in the figures below satisfy these conditions. The one on the left is made only of triangles, with every vertex having a valency of 6. The one on the right is made only of quadrilaterals, with every vertex having a valency of 4.

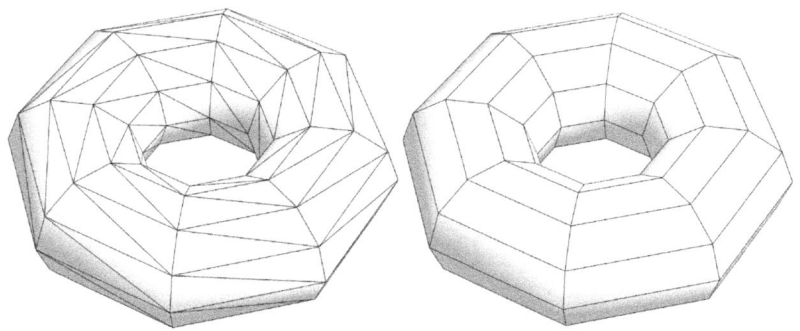

Figure 13.6 A polyhedron in the shape of a donut.

This shows that by studying how a surface is divided into polygons, we can figure out whether the shape is like a sphere or like a donut. In other words, we don't have to look at the shape from the outside. If we could move around on the surface like an ant, we could still determine its overall form!

CHAPTER 14

Success Rate of the "Kendama" Challenge

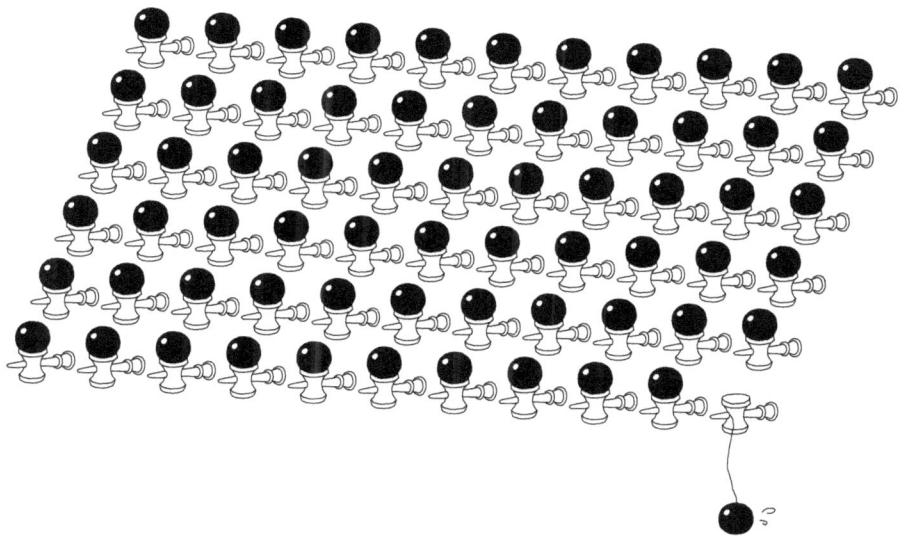

In recent years, the NHK Kōhaku Uta Gassen—a popular year-end music show broadcast throughout Japan—has included a unique segment: a group attempt to break a Guinness World Record using the traditional Japanese cup-and-ball toy *kendama*.[1] This segment involves participants

---

[1]Kendama is a traditional Japanese toy and performance skill game. It has a wooden handle shaped like a hammer with three cups (two on the sides and one at the bottom) and a spike at the top. A wooden ball with a hole is attached to the handle by a string. The main goal is to catch the ball in the cups or to land the hole

DOI: 10.1201/9781003670261-14

performing a basic kendama trick known as *ōzara*, or "big cup," where the ball is caught on the larger of the two side cups of the toy. Each person takes a turn, and the goal is to count how many people can successfully complete the trick in a row.

At the time of writing, the standing world record is 127 consecutive successful attempts. During the 2023 edition of the show, a new record was attempted with 128 participants. Unfortunately, the challenge fell short of success.[2] Given that the participants are likely quite skilled in kendama, this result highlights just how much pressure can come from performing under the intense spotlight of national television.

A MATTER OF PROBABILITY

Suppose that each participant has a 99% chance of successfully performing the *ōzara* trick. What, then, is the probability that all 128 participants will succeed in a row?

This can be calculated simply by raising 0.99 to the 128th power:

$$0.99^{128} = 0.276\ldots$$

From this result, we see that the probability of all 128 participants succeeding is less than 28%. Even if every person has a 99% success rate, the chance of setting a new record drops below 28%. This really shows just how tough the challenge is.

Now, let's consider the opposite: if we want at least a 50% chance of setting a new record, what individual success rate would each participant need to achieve? After all, if you're organizing such an event, you'd probably want better than even odds of success.

We can find this by solving the following equation, where x is the success rate of each individual:

$$x^{128} = 0.5.$$

To solve this, we take the logarithm of both sides. (The base of the logarithm doesn't matter for this calculation.)

---

of the ball on the spike. Players can also perform many tricks by combining catches, swings, and balances.

[2] In the 2024 challenge, however, the participants succeeded spectacularly.

$$\log x^{128} = \log \tfrac{1}{2}$$
$$128 \log x = \log \tfrac{1}{2}$$
$$\log x = \tfrac{1}{128} \log \tfrac{1}{2}$$
$$\log x = \log \left(\left(\tfrac{1}{2}\right)^{\frac{1}{128}} \right)$$
$$x = \left(\tfrac{1}{2}\right)^{\frac{1}{128}} = 2^{-\frac{1}{128}} \approx 0.99459\ldots$$

From this, we find that each participant would need a success rate of approximately 99.5%. (This is a great example of how logarithms—something taught in high school—can be useful for solving real-world problems.) Even so, this reveals just how tough the challenge truly is.

Now, what if we allowed for one mistake among the 128 attempts? Suppose that if someone fails, they are given one extra try. However, this allowance can only be used once for the entire group.

The probability of success under this condition can be calculated using the following formula:

$$0.99^{128} + 128 \times 0.01 \times 0.99^{128}.$$

The first part, 0.99^{128}, represents the probability that all participants succeed on their first try. The second part considers the scenario where one person fails initially but succeeds on their second attempt—this has a probability of 0.01×0.99. The remaining 127 participants must still succeed on their first try, which contributes a factor of 0.99^{127}. Since any one of the 128 participants could be the one who makes the mistake, we multiply this by 128. When we calculated this, it came to about 63%.

With just one allowed mistake, a 99% success rate per person gives the group more than a 60% chance of succeeding. That's a noticeable improvement.

Now, let's raise the bar: suppose each participant has an extremely high 99.9% success rate. In this case, what is the maximum number of participants n such that the group still has a 50% chance of success? We need to solve the following equation:

$$0.999^n = 0.5.$$

This is another exponential and logarithmic problem. We solve it by taking the logarithm of both sides:

$$\log 0.999^n = \log \tfrac{1}{2}$$
$$n \log 0.999 = \log \tfrac{1}{2}$$
$$n = \frac{\log \tfrac{1}{2}}{\log 0.999} \approx 692.8 \ldots .$$

If each participant could perform the *ōzara* trick with a 99.9% success rate (meaning only one mistake is allowed in every 1,000 tries!), the calculation shows that a record of 692 people could be achieved with a 50% chance of success.

Of course, as mentioned at the beginning, the participants are likely under immense pressure, so things aren't nearly as simple as the math suggests. Calculations are just that—calculations. The real world is far more complex.

By the way, if you're like me and tend to fail about once every 10 attempts, the probability of succeeding 128 times in a row is

$$0.9^{128} = 0.00000139 \ldots$$

That's pretty much hopeless, wouldn't you say?

CHAPTER 15

Estimating the Dimensions of a 1-Liter Milk Carton

Figure 15.1　1-Liter milk carton.

Milk is one of the essential beverages in our daily lives, but the way it's sold varies from country to country. In Japan, it's commonly sold in paper cartons that hold 1 liter. These paper cartons are typically shaped like rectangular prisms, as shown in the illustration above.

DOI: 10.1201/9781003670261-15

Now, here's a sudden quiz: How wide is this paper carton in centimeters?

If you've never measured one before, you might not be able to answer right away. But using the clue that the volume is 1 liter, or 1000 mL, let's try estimating its approximate dimensions. Using the method I'm about to explain, I was able to guess the exact answer.

ESTIMATING THE SIZE OF A MILK CARTON

Take a good look at the shape of the milk carton shown at the beginning. Try estimating its width. It's unlikely that the milk fills all the way up to the roof-like top of the carton, so we can ignore that part and focus on the rectangular prism portion. We can assume that the volume of this rectangular section is 1000 mL, or 1000 cm^3.

That means, using centimeters as the unit, the width, depth, and height of the milk carton should satisfy:

$$(\text{width}) \times (\text{depth}) \times (\text{height}) = 1000.$$

Next, let's make a simplifying assumption—even if it's not perfectly accurate—that the base of the carton is a square. In other words,

$$(\text{width}) = (\text{depth}).$$

From how the carton looks from the front, the height seems to be about 2 to 3 times the width, so let's estimate it as roughly 2.5 times the width.

With these assumptions, if we let the width be x, then the depth is also x, and the height is $2.5x$. This gives the volume as:

$$x \times x \times 2.5x = 2.5x^3.$$

If we assume that this value equals 1000, then:

$$2.5x^3 = 1000$$
$$x^3 = 400$$
$$x = \sqrt[3]{400}.$$

Now, it's hard to calculate the exact value of x without a calculator, but let's try estimating it.

We can break down 400 as:

$$400 = 2 \times 2 \times 2 \times 2 \times 5 \times 5.$$

So,

$$x = \sqrt[3]{400} = 2\sqrt[3]{50}.$$

We're almost there.

To estimate the value of $\sqrt[3]{50}$—the cube root of 50—without using a calculator, we can recall that:

$$3^3 = 27, \quad \left(\sqrt[3]{50}\right)^3 = 50, \quad 4^3 = 64.$$

So, $\sqrt[3]{50}$ must be somewhere between 3 and 4. Let's estimate it as roughly 3.5.

From earlier, we had:

$$x = 2\sqrt[3]{50}.$$

Substituting in our estimate:

$$x \approx 2 \times 3.5 = 7.$$

With this reasoning, I estimated that the width of the milk carton is about 7 cm.

Now then, what's the actual value...?

When I actually measured the width of a milk carton, it was exactly 7.0 cm—much to my own surprise!

As you can see, even what might seem like a rough estimate can sometimes lead you to the correct value. Estimating the dimensions of various objects is a fun way to sharpen your intuition. Try guessing the sizes of everyday items around you!

By the way, the actual dimensions of a 1-liter milk carton are said to be 7.0 cm in both width and depth, and 19.5 cm in height. If we calculate the volume using these values:

$$7 \times 7 \times 19.5 = 955.5 \text{ mL}$$

—that's less than 1000 mL.

This discrepancy is due to a clever design feature: the carton slightly bulges in the middle when filled with milk, which increases the volume. So even if the surface area remains the same, the volume can change depending on the shape.

Even a single milk carton reveals surprising depth and insight!

THE NET OF A MILK CARTON

The spout area of a milk carton has a roof-like shape. Looking at the circled area in the net diagram below, you can see that this shape is formed by folding part of the side of the rectangular prism into a triangular shape.

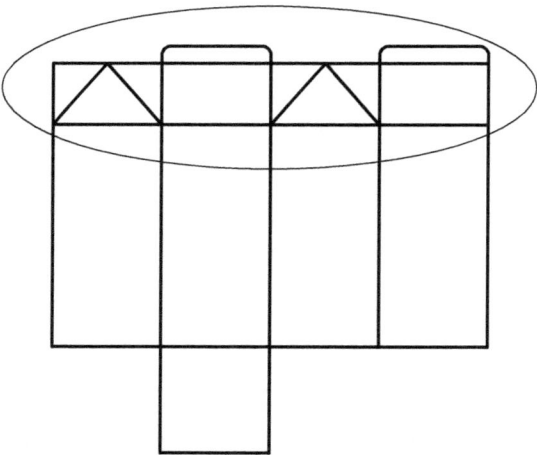

Figure 15.2 The net of a 1-liter milk carton.

Since there are no cuts made in the material, it looks like the top portion of the rectangular prism has been folded using origami.

Wondering if I could give this roof part a more stylish design, I created my own origami version. It's shown in the figure below, and I named it "Tulips on Top of Milk Cartons".

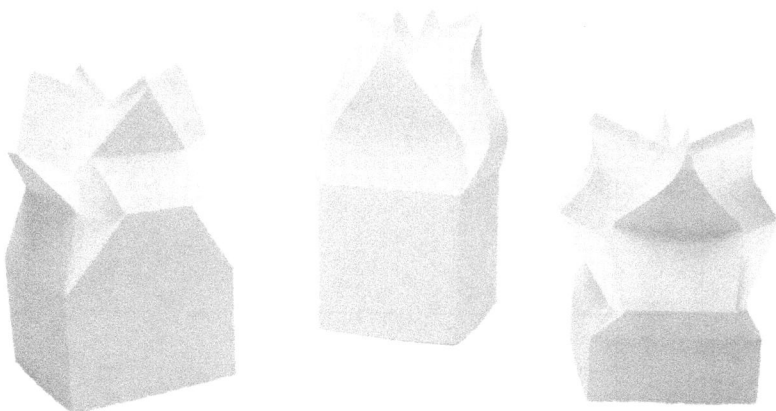

Figure 15.3 Creation: Tulips on top of milk cartons.

Figures below show their net (if you want to actually make it, you'll need to add a glue flap on either the left or right side so that it can be assembled into a cylinder).

They are not practical at all, but I was able to create cute shapes using just part of a rectangular prism.

In the upcoming section "Chapter 32: How Can We Fit As Much As Possible into an Envelope?" we'll also talk about how different shapes can have different volumes even when the surface area stays the same. Stay tuned!

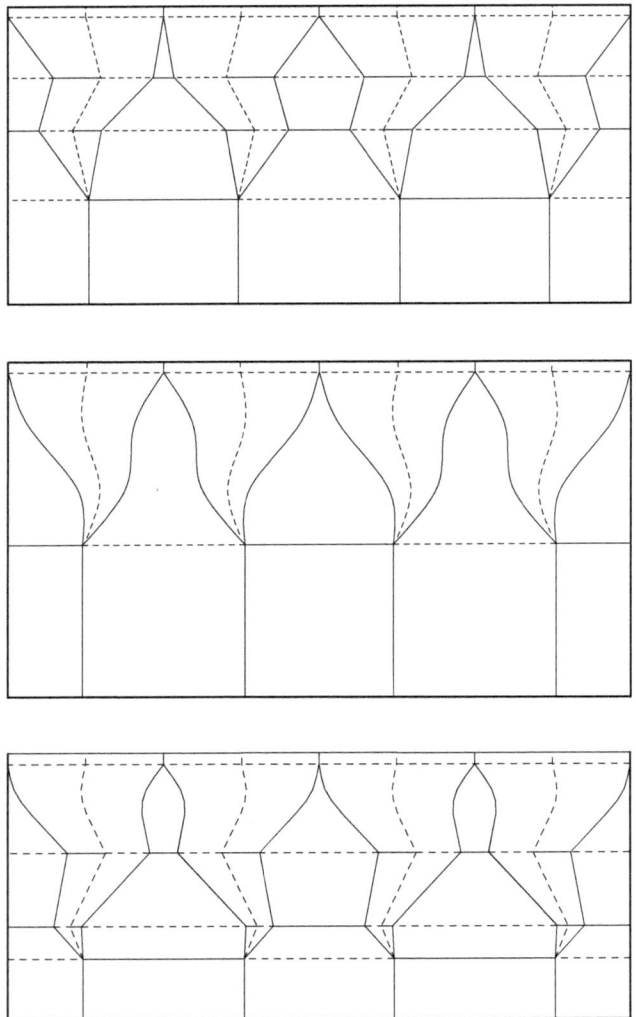

Figure 15.4 Nets of the creations.

CHAPTER 16

Seats on the Shinkansen (Japan's bullet train) Divided into Two and Three Rows

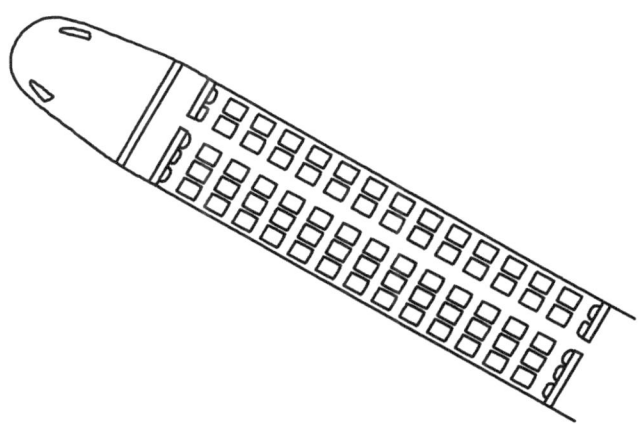

Figure 16.1 Seating arrangement on the Shinkansen.

On the "Shinkansen" (Japan's bullet train), the seating is divided by an aisle: one side has two seats, and the other side has three. Although this arrangement isn't symmetrical, it actually works out quite well when traveling in groups. Regardless of the group size, it's possible to sit to-gether neatly without leaving anyone alone.

For example, a group of four can use two two-seaters. A group of five can take one two-seater and one three-seater, allowing everyone to sit together without anyone being left out. For six people, two three-seaters work perfectly. For seven, two two-seaters and one three-seater do the trick.

WHY THE SEATING WORKS SO WELL

No matter the group size (as long as it's at least two), it's always possible to sit without leaving any seats unused. Some may intuitively understand why this is the case, but others might need more convincing. So, let's go ahead and prove it.

We'll start by dividing the cases based on whether the number of people is even or odd.

(**Proof 1**) Let n be the number of people (where $n \geq 2$). If n is even, we can use only two-seaters to seat everyone without any leftover seats. If n is odd, we first seat three people in a three-seater. That leaves $n - 3$ people. Since $n - 3$ is even, we can use only two-seaters for the rest. $Q.E.D.$

Next, let's use mathematical induction to prove the same idea.

(**Proof 2**)

1. For $n = 2$, the two people can sit in one two-seater.

2. Assume that for $n = k$, the people can sit without any unused seats. Then, for $n = k + 1$: If two of the seated people are in a two-seater, we can move them to a three-seater and add one more person to that seat. If no one is in a two-seater, we must have at least one three-seater in use. We can move three people from that seat and seat them as two pairs in two-seaters, making room for the extra person.

Therefore, by (1) and (2), this seating rule holds for all $n \geq 2$. $Q.E.D.$

This may seem like a complicated explanation, but it's actually a great example for practicing how to write a mathematical proof. It also shows that there's more than one way to prove the same idea.

From a practical standpoint, though, if everyone tries to use only two-seaters, the three-seaters will end up mostly empty. So how can we make good use of both types of seats?

(**A Balanced Way to Use Two-Seaters and Three-Seaters**) Let x be the remainder when the number of people is divided by 5. Seat people starting from the front of the train and from the end of the two-seater side, as shown in the diagram.

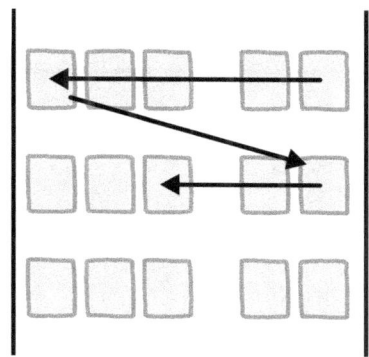

Figure 16.2   Order of seating.

- If $x = 0$, everyone fits perfectly.

- If $x = 2$, the remaining two people sit in a two-seater.

- If $x = 3$, the remaining three people sit in a three-seater.

- If $x = 4$, the remaining four people sit in two two-seaters.

- If $x = 1$, move one pair from a two-seater to a three-seater, and seat the remaining person in the now-available seat.

Using this method, you can fill seats from the front and seat everyone efficiently without leaving any extras.

This step-by-step method is like a program that gives instructions to a computer. A predefined sequence like this is called an *algorithm*. Thinking algorithmically in everyday life can help you do things more efficiently.

WHERE WOULD YOU SIT IN A THREE-SEATER?

If you had to choose a seat in a three-person row, where would you prefer to sit?

Many people might go for the window seat, where you can enjoy the view. Others might prefer the aisle seat so they won't disturb anyone when they need to get up and use the restroom.

Interestingly, perhaps anticipating these preferences, the middle seat in a three-seater is actually made slightly wider—just a few centimeters—to offer a bit more comfort. So, would you choose the view, the ease of getting up, or the extra space? It's a tough decision!

II

Playful Math You Can Touch and Feel

Crafting is full of mathematical elements. Try arranging building blocks, or folding and cutting familiar materials like paper or cardboard, then combining them in different ways. You'll be able to *feel* the beauty of the mathematical world right in the palm of your hand.

CHAPTER 17

Enjoying the Cleanup of Building Blocks

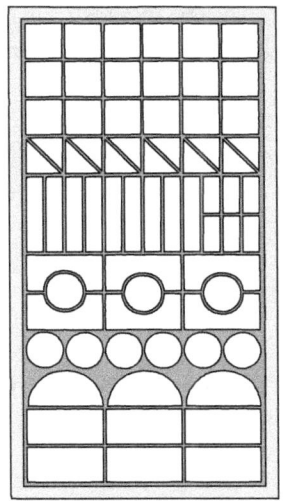

Figure 17.1 Different ways to arrange blocks in a box.

Cleaning up building blocks after playing with them can feel like a chore—but if you try out different ways to pack them, you might discover something surprisingly fun. Instead of lining up the blocks with the edges of the box as usual, you can try arranging them diagonally, like on the right side of the figure above. You might be surprised how neatly they fit.

DOI: 10.1201/9781003670261-17

THE JOY OF BUILDING BLOCKS

Building blocks are a classic toy for young children. Stacking various shapes to create castles, houses, or bridges is loads of fun. At the same time, kids naturally learn about geometric solids like cubes, rectangular prisms, cylinders, and triangular prisms. Even if you're not constructing anything specific, you can enjoy stacking blocks as high as possible or simply arranging them in neat rows. Knocking them down is also part of the fun. It's easy to see why building blocks are beloved all over the world.

But what if I told you that the joy of building blocks extends even to *cleaning them up*? That might sound surprising. It's common to leave the blocks scattered, get scolded, and then reluctantly put them away. However, when you carefully arrange them like in the figure above, you'll start to see geometric tiling patterns emerge. Most building blocks are simple shapes like rectangular prisms or cylinders, and in the case of rectangular prisms, their sides often follow easy-to-work-with ratios like 1:2:4. Thinking about how to neatly pack them into a wooden (or paper) box can be quite engaging.

Usually, cubes and rectangular prisms are placed parallel or perpendicular to the box edges. But the blocks I have at home, just like in the figure above, fit perfectly even when some of them are tilted at a 45° angle. That was an unexpected discovery. Why does that work so well?

DIMENSIONS OF THE SHAPES

In the arrangement on the left side of the figure at the beginning, six cubes fit perfectly side by side in the wooden box. If we let the side length of each cube be 1, then the width of the box is 6.

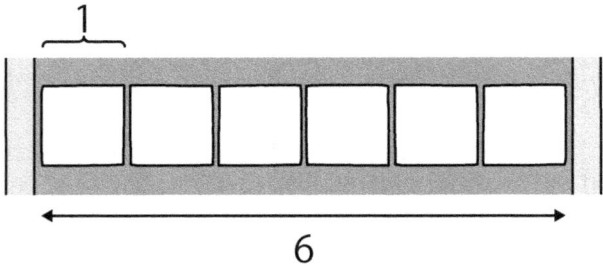

Figure 17.2 Six cubes arranged in a line.

Now, what happens in the arrangement on the right side of the figure, where we place a rectangular prism with a width of $\frac{1}{2}$, followed by four cubes tilted at a 45° angle? The diagonal of a square with side length 1 is $\sqrt{2}$ (approximately 1.414), so the total width of this arrangement, as shown in the diagram below, is $0.5 + \sqrt{2} \times 4 \approx 6.157$.

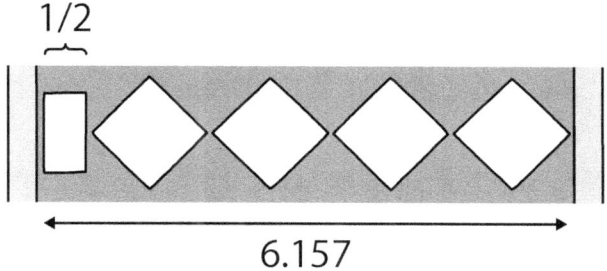

Figure 17.3 Tilted cubes arranged in a row.

The total width is slightly larger than in the earlier case. Still, everything fit just fine because the wooden box was made with a bit of extra room. If the box had been sized exactly, it would have been difficult to take the blocks in and out. Thanks to this small allowance for smooth handling, such an unconventional (yet beautiful) packing method became possible.

The width of the tilted arrangement, 6.157, is about 2.6% greater than the original width of 6. Since there's still a little space left even in the tilted setup, we can see that the box was originally designed with a few percent of leeway.

THE DENSEST PACKING OF SHAPES

Now, let's turn our attention to the arrangement of cylinders on the right side of the first figure. It looks like the circles (representing the cylinder bases) are packed quite efficiently. When it comes to arranging circles in a regular pattern, two common methods quickly come to mind, as shown in the following diagram.

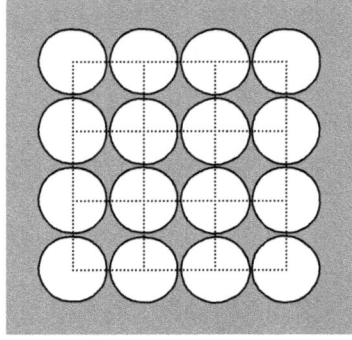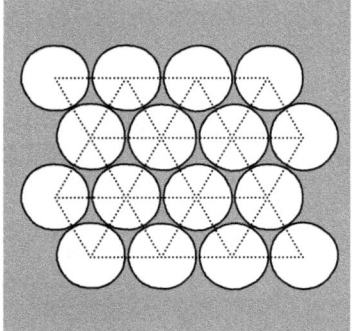

Figure 17.4 Ways to tile with circles.

On the left, the circles are arranged so that their centers form a square grid. On the right, the centers form a pattern of equilateral triangles. This method reduces the empty space between the circles.

If you're packing circles across an unlimited area, the method on the right is known to be the most efficient. It allows about 15.6% more circles to be packed than the left-side method. The proportion of the area occupied by the circles is called the *packing density*. In the left-side arrangement, the packing density is 78.5%, while on the right it is 90.75%. That means the empty space is 21.5% on the left and only 9.25% on the right. Clearly, the arrangement on the right is much more efficient.

But this efficiency assumes that you're packing a very simple shape—circles—and that the area is large enough. If the shapes become more complex, if you're dealing with a mix of different shapes, or if the area you're trying to fill is irregular, then the problem of "how to pack the shapes with the least empty space"—known as the *densest packing problem*—becomes much harder.

Thinking about how to efficiently clean up and store building blocks connects directly to these kinds of challenging mathematical problems.

Today we've looked at packing shapes in a flat plane. When it comes to packing 3D shapes into space, the problem is just as important—whether you're studying the arrangement of atoms and crystal structures in science, or figuring out how to pack everyday objects. Think about those "fill-the-bag for 1,000 yen" deals—you definitely want to pack things in as tightly as you can!

DENSEST SPHERE PACKING

To pack spheres of the same size most efficiently in three-dimensional space, the best approach is the way of stacking as shown on the right side of figure below. You start by arranging the first layer of spheres in triangular lattice form, then place the second layer so that each sphere fits into the hollow formed by three touching spheres from the layer below. This arrangement, found in various crystal structures, is known as *hexagonal close packing* (there is also an alternative method called *face-centered cubic packing*, which achieves the same density).

Although this sphere arrangement had long been believed to be the most efficient, it wasn't until surprisingly recently—1998—that it was actually proven to be true.[1]

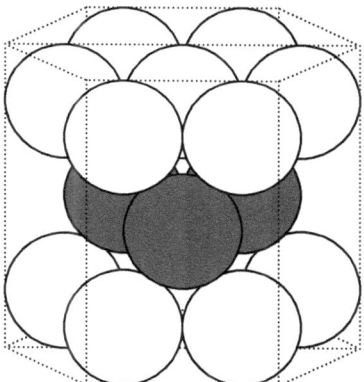

Figure 17.5   Hexagonal close packing of spheres.

---

[1]In 1998, Thomas C. Hales confirmed it using a method originally proposed by László Fejes Tóth.

CHAPTER 18

The Mysterious Surface Made of Cotton Swabs: A Hyperboloid

Figure 18.1   Tilted cotton swabs seen inside a transparent container.

Cotton swabs are often sold in cylindrical transparent containers. As you take them out one by one, a beautiful shape like the one in the photo above may appear. Have you ever seen it? This shape is known as a *hyperbolic paraboloid*.

                                    DOI: 10.1201/9781003670261-18

SURFACES MADE OF STRAIGHT LINES

When you hear the word "surface," you might picture something smooth and curved, like the surface of a sphere or the body of a car—shapes without sharp edges. So, it might seem surprising to hear that there are surfaces made entirely of straight lines.

These are called *ruled surfaces*, and they are formed by tracing the path of a straight line as it moves through space. In other words, they are surfaces constructed from straight lines.

Familiar examples include the side surfaces of *cones* and *cylinders*. As shown in the figures below, these shapes can be represented as a collection of straight lines.

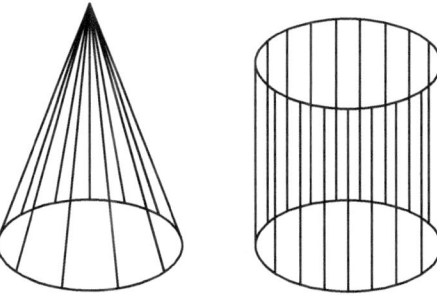

Figure 18.2 Ruled surfaces (cone and cylinder).

In mathematics, cones and cylinders are understood in a slightly more general way. A surface formed by straight lines that all pass through a single point is called a *conical surface*, while one formed by parallel straight lines is called a *cylindrical surface*.

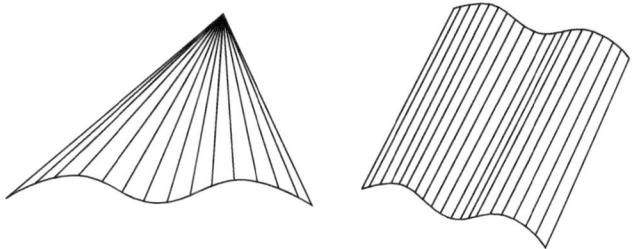

Figure 18.3 Ruled surfaces (conical and cylindrical surfaces).

There are many other types of ruled surfaces as well. The diagram below shows a shape called a **hyperbolic paraboloid**, which is indeed made up of a collection of straight lines.

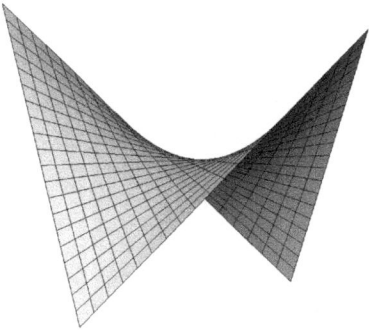

Figure 18.4 Hyperbolic paraboloid.

This surface has another fascinating property. When it is sliced by a horizontal plane, the cross-section forms a *hyperbola*, as shown on the left in the figure below. When it is sliced by a vertical plane, the cross-section becomes a *parabola*. This clearly explains why it is called a hyperbolic paraboloid.

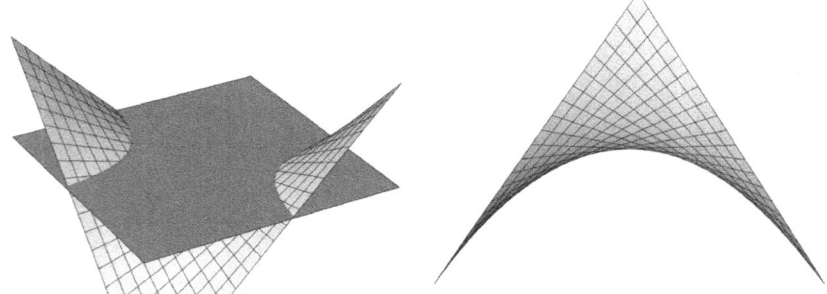

Figure 18.5 Cross-sections of a hyperbolic paraboloid.

Surfaces made from straight lines like this are well-suited for architecture. After all, you can create a curved surface simply by assembling straight structural elements. The photo below shows a structure located near a park in the author's hometown. The shape of its roof is a hyperbolic paraboloid.

Figure 18.6　A structure shaped a hyperbolic paraboloid.

COTTON SWABS AND THE HYPERBOLOID

Many of you probably have cotton swabs at home. They are usually sold in cylindrical containers, and as you take them out one by one, a beautiful form like the one shown in the opening photo sometimes appears when only a few dozens are left.

Interestingly, it's quite difficult to create this shape by hand starting from an empty container, but if you simply remove the swabs from a full container, the shape naturally forms. It's quite a mystery, isn't it?

The shape formed by these tilted cotton swabs shares the characteristics of a surface called a *hyperboloid* or *hyperboloid of one sheet*. As shown in the diagram below, imagine two circles placed parallel to each other—one above the other. Then, draw straight lines connecting pairs of points moving at the same speed along the circumference of each circle. The key is to slightly offset the starting positions of the points on the top and bottom circles. Doing so reveals the shape of a hyperboloid.

This surface has distinctive features: when sliced with a horizontal plane, the cross-section is a circle; when sliced with a vertical plane, the cross-section becomes a hyperbola.

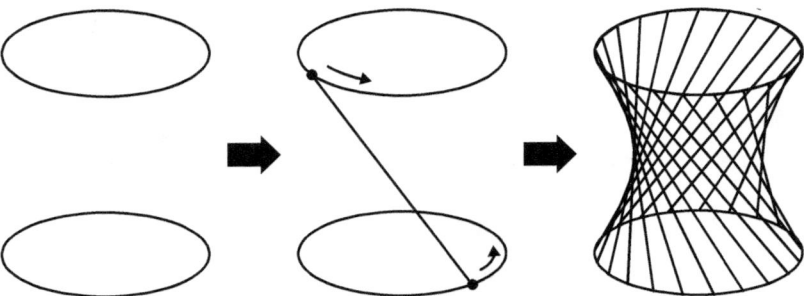

Figure 18.7 Constructing a hyperboloid.

The diagram above shows the same structure formed by the slightly tilted cotton swabs inside a cylindrical container. It's truly enjoyable to encounter the world of geometry through such everyday items.

Now, how likely is it that you'll actually observe this cotton swab hyperboloid? Suppose the shape tends to appear when there are around 20 to 40 swabs left. If you purchased a pack of 200, then the chance that this number of swabs remains is about 10%. Furthermore, the likelihood that you've managed to remove the swabs up to that point without disrupting the shape might be around 50-50. That means about 5% of households with cylindrical cotton swab containers might get to observe a hyperboloid of one sheet. If you have a cylindrical container of cotton swabs at home, why not check what shape they're forming right now? (This kind of estimation, using reasonable assumptions to arrive at an approximate value when a direct measurement is difficult, is known as a *Fermi estimate*.[1])

By the way, what determines the direction of the twist? When I observed it, the swabs were tilted counterclockwise when viewed from above. (I even confirmed that the swabs at my parents' house during a visit also tilted counterclockwise.) Does it ever twist clockwise instead? Could it have something to do with whether you're right- or left-handed?

---

[1] Named after the physicist Enrico Fermi, who was known for his ability to make such approximations.

COTTON SWAB CRAFT

The hyperboloid formed by cotton swabs is so fascinating that I decided to glue them together with adhesive and take it out as a solid model.

Figure 18.8 Glued and taken out from the container.

Cotton swabs make excellent materials for crafting geometric shapes. They offer the advantage of being uniform in length and easy to glue together at the tips, making them perfect for model construction. For example, as shown in the photo below, you can build a regular icosahedron using just 30 cotton swabs.

Figure 18.9 Crafting with cotton swabs.

You don't have to use cotton swabs to create the shape of a hyperboloid. Any thin, stick-like material will do, and toothpicks—something found in almost every household—work just as well. In the photo below, I used a 3D printer to make the base supports and then formed the curved surface using toothpicks. Crafting with everyday items like this can be a great way to refresh your mind.

Figure 18.10 Hyperboloid made with toothpicks.

ONE STEP FURTHER

When cotton swabs form a hyperboloid, it appears that they are supporting each other inside the container. If that's the case, then the final shape should depend on factors such as the length, thickness, and number of swabs, as well as the diameter of the container. These values that determine the final shape are called *parameters*.

By changing these parameters, we can use computer graphics to simulate the different forms that can be produced.

By observing cotton swabs, we can learn a great deal about the geometry of surfaces and curves.

Figure 18.11 Various hyperboloids recreated with computer graphics.

CURVE FORMED BY THE INTERSECTION OF A SPHERE AND A CYLINDER

The following figure simulates the trajectory drawn when a cotton swab gradually tilts while both ends remain in contact with the container.

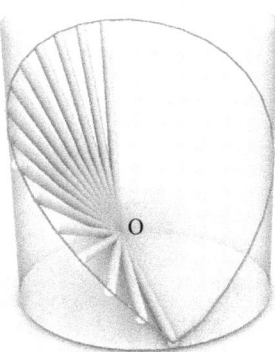

Figure 18.12 Curve traced by the tip of a cotton swab as it falls inside the container.

Let's fix one end of the cotton swab and call this point O. The opposite end will then move along the surface of a sphere centered at point O, with a radius equal to the length of the swab. However, since the end of the swab is also touching the cylindrical container, the curve shown in the diagram represents *the intersection of a sphere and a cylinder.*

When the diameter of the container matches the length of the swab, the resulting curve is known as the *Viviani curve.*[2]

---

[2]Named after the Italian mathematician Vincenzo Viviani.

CHAPTER 19

With 30 Toy Train Tracks, You Can Play for Over 200 Years

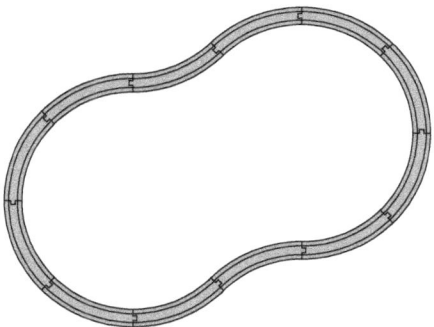

Figure 19.1 Examples of plarail layouts.

Plarail is a popular Japanese toy train system that allows you to create a wide variety of layouts by combining straight tracks, curved tracks, switches, and more. At the time of writing, Plarail is celebrating its 65th anniversary.[1] With such a long history, many people likely have memories of playing with it across generations.

Even if you have only a limited number of track pieces, you can still

---

[1]Believe it or not, I had the honor of serving as an official ambassador for Plarail's 65th anniversary.

DOI: 10.1201/9781003670261-19

come up with many different layouts just by changing how you arrange them.

MODEL TRAIN TOYS

If you've ever played with a model train set like Plarail, you may have found yourself wondering: "What are all the possible layouts I can make with just the tracks I have?" Even with a limited number of pieces, the way you arrange them can result in completely different track shapes.

For example, suppose you only have three types of track: straight, right curve, and left curve. With just two pieces, you already have $3 \times 3 = 9$ combinations. With three pieces, it's $3 \times 3 \times 3 = 27$, and with ten, there are a staggering $3^{10} = 59{,}049$ possible arrangements.

Now, let's narrow it down and count only the layouts where the train loops back to its starting point. When it comes to the tedious task of listing *every possible combination that meets certain conditions*, computers are your best friend. So let's try investigating what kinds of layouts can be created under the following conditions:

- Combine straight tracks and curved tracks (each curved track turns one-eighth of a full circle) to form a closed loop that returns to the starting point.

- Layouts are allowed to cross over themselves.

- If two layouts are the same after rotating, flipping, or reversing the direction of travel, we consider them identical.

At first glance, listing out all these conditions might feel like a hassle, but when you're trying to enumerate patterns systematically, it's a necessary step.

Now, let's take a look at the computer-generated results, sorted by the total number of track pieces used. By the way, the number of pieces needed to make a closed loop must be an even number, and at least 8 (we'll skip the proof, but it's fairly intuitive that you need at least that many).

NUMBER OF LAYOUTS YOU CAN MAKE WITH A LIMITED NUMBER OF TRACKS

8 Pieces

The simplest closed layout you can make uses just 8 curved tracks, forming a perfect circle, as shown in the figure below. It's easy to see that with fewer pieces than this, creating a closed loop is impossible.

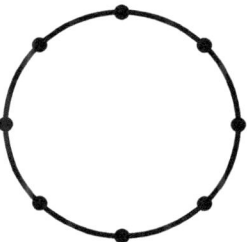

10 Pieces

The next simplest layout uses 8 curved tracks combined with 2 straight tracks, as shown in the figure below. Again, there is only one possible layout with this configuration.

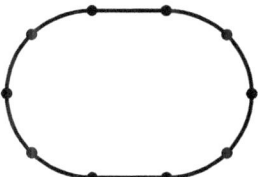

12 Pieces

Next is the case with 12 pieces. As shown in the figures below, there are a total of 4 distinct layouts: one made entirely of curved tracks (far left), and three others that include 4 straight tracks.

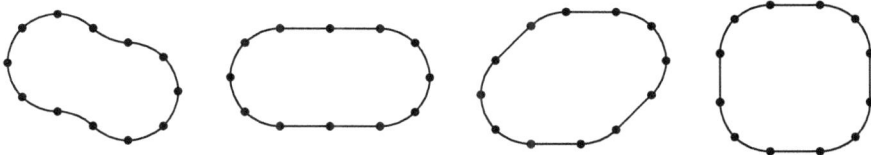

14 Pieces

When using 14 pieces, there are 9 different layouts you can create, as shown in the figures below. One of them, the heart-shaped layout located second from the left in the bottom row, has a distinctive design that may not be easy to come up with, but it is well known among fans of Plarail.

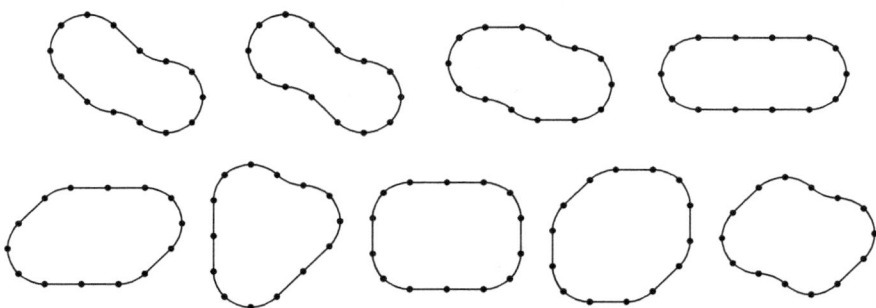

16 Pieces

With 16 pieces, there are 42 different layouts, as shown in the figures below. One of them, indicated by arrows, includes a full circle made of 8 curved tracks. This means the layout loops around the circle twice, which isn't practical for an actual train to run on.

Now, since it becomes difficult to show every layout visually beyond this point, let's just look at the numbers.

As the number of track pieces increases, the number of possible layouts grows dramatically. This is a clear example of *combinatorial explosion*.

Having around 30 tracks is not unusual for a typical Plarail set. But as the table shows, there are over 9.7 million different layouts possible with 30 pieces. If you were to build and play with 100 different layouts every day, it would take you 268 years to go through them all. That's an almost unimaginable number.

However, it's important to note a key detail here. In this count, the breakdown between curved and straight tracks was not fixed. That means layouts using 10 straight and 20 curved tracks are included, as well as those using only 30 curved tracks. If you limit the count to specific combinations of track types, the number of possible layouts becomes much smaller.

DRAWING GEOMETRIC SHAPES WITH PLARAIL

Plarail is designed to let you enjoy running toy trains, but if you think of straight tracks as line segments and curved tracks as arcs of a circle, the layouts you build can also be seen as geometric figures composed only of lines and arcs.

So, what kinds of shapes can you actually draw? This is another case where using a computer makes it easy to explore.

Table 19.1 Number of possible layouts by track count

Number of Tracks	Number of Possible Layouts
18	161
20	847
22	4,739
24	29,983
26	198,683
28	1,375,928
30	9,786,630

The figure below shows an overlay of all the possible shapes that can be formed by connecting just two pieces of track. Starting from the bottom (assuming you're moving upward from the starting point), you have three options: go straight, curve left, or curve right. From each of those positions, you can again choose from the same three options. This results in a branching pattern that leads to a total of nine different endpoints.

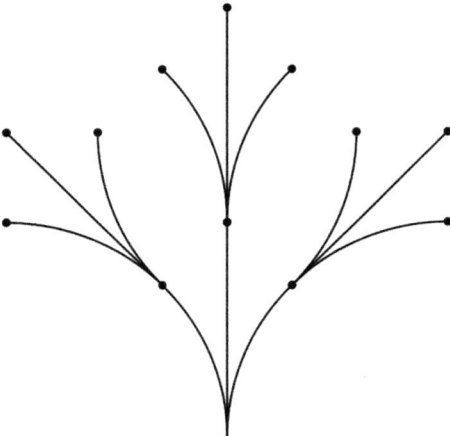

Figure 19.2 Overlay of shapes formed by all combinations of two track pieces.

What happens if we add one more piece of track? Each of the nine endpoints branches out into three new directions, resulting in the tree-like structure shown in the figure below.

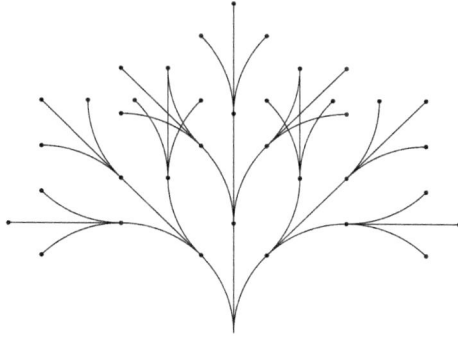

Figure 19.3 Overlay of shapes formed by all combinations of three track pieces.

If you look closely, you'll notice that some paths actually converge. For example, the sequence of straight → right curve → left curve ends at the same position and with the same direction as the sequence of right curve → left curve → straight. In other words, although the order of the pieces is different, the final position and orientation of the train are identical.

Let's keep going and increase the number of track pieces in the same way. The figures below show tree structures created using 4 pieces (left) and 5 pieces (right). As you can see, the branches multiply rapidly. With 5 pieces, there are $3^5 = 243$ possible layouts included.

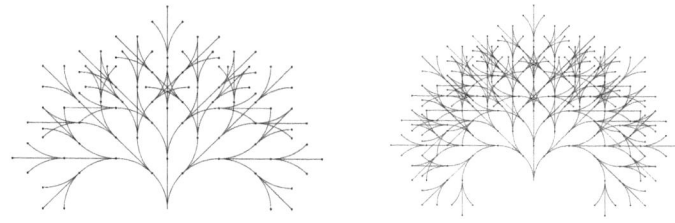

Figure 19.4 Overlay of shapes formed by all combinations of 4 (left) and 5 (right) track pieces.

Now let's take things a step further. Suppose we don't limit the initial direction to straight upward but instead allow movement in any of

Figure 19.5 Overlay of shapes formed by all combinations of 4 (left) and 12 (right) track pieces.

the eight directions—up, down, left, right, and the four diagonals. What kind of patterns can we create?

The left side of the figures below shows the pattern that can be made with 4 track pieces. The right side shows a pattern formed by combining 12 curved tracks only. The result looks almost like a *mandala*.

Exploring the geometric shapes that can be drawn with Plarail reveals a world full of interesting discoveries.

COORDINATES OF TRACK ENDS

The diagram below shows only the endpoints of tracks from patterns created using 7 pieces, with the connecting lines removed. It looks like there might be some kind of pattern hidden in the arrangement.

It turns out that the coordinates of these points can be expressed using the following formulas:

$$\left(m_x + \frac{1}{\sqrt{2}}n_x, m_y + \frac{1}{\sqrt{2}}n_y\right).$$

Here, m_x, n_x, m_y, n_y are integers that satisfy the condition $n_x \pm n_y \equiv 0$ (mod 2).

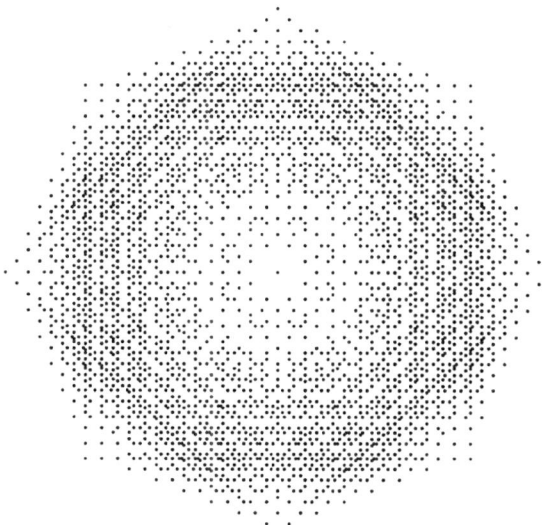

Figure 19.6 Track ends of shapes formed by combinations of seven track pieces.

While playing with Plarail alongside your child, you might find yourself enjoying some surprisingly deep mathematical exploration.

CHAPTER 20

Is It Easy or Difficult to Make a Plarail Layout That Loops Back to the Start?

Figure 20.1 Geometric patterns made with tracks.

DOI: 10.1201/9781003670261-20

With Plarail layouts, you can create symmetrical and beautiful geometric patterns. The method is surprisingly simple: just take any sequence of tracks and repeat it over and over. Eventually, the ends will connect, forming a continuous loop like the one-stroke patterns shown in the figure above.

Of course, the tracks may cross over each other, so you can't actually run a train on them. But it's still fun to try, so give it a go!

GEOMETRIC PATTERNS MADE WITH PLARAIL TRACKS

The geometric pattern shown in the opening figure is created by repeating the track sequence in the diagram below 8 times in a row. By simply repeating the same shape 8 times, the end of the track returns exactly to the starting point, allowing the beginning and end to connect perfectly.

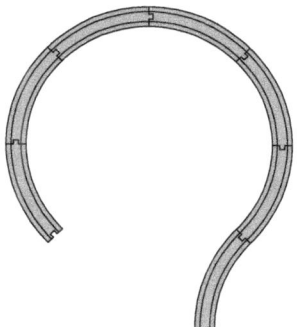

Figure 20.2   Arrangement of tracks.

The reason you can create such beautiful patterns is not because the specific sequence of tracks in the figure above is special. In fact, almost any sequence will work. By repeating it 2, 4, or 8 times, the track eventually loops back to its starting point, forming a closed layout.

The only exception is a sequence where the direction at the start and end is exactly the same, such as simply arranging straight tracks in a row. In that case, the track will just keep extending farther away.[1]

---

[1]But then again, who knows—maybe one day the tracks could loop all the way around the Earth and connect back together.

Other examples are shown in the next figure. The layout on the left returns to the starting point after 2 repetitions, while the one on the right does so after 4 repetitions.

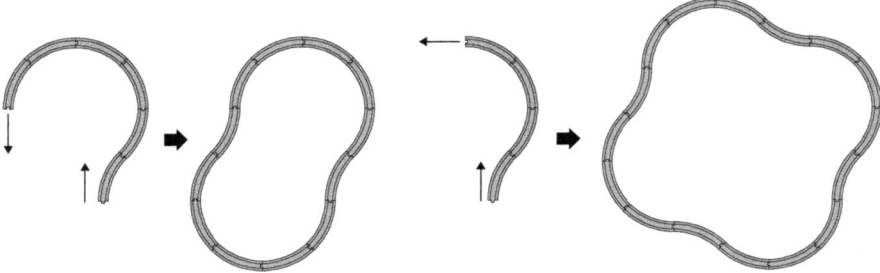

Figure 20.3 Example that returns to the starting point after 2 repetitions (left) and after 4 repetitions (right).

In the figure, arrows are used to indicate the directions of travel at the starting and ending points for clarity. The number of repetitions needed depends on the difference in angle between these two arrows, as shown in the following table.

Table 20.1 Angle Difference and Number of Repetitions

Angle difference	45°	90°	135°	180°
Number of repetitions	8	4	8	2

By using this property, you can create a wide variety of geometric patterns that eventually loop back to the starting point. However, in some cases the tracks may intersect with each other. When that happens, trains cannot actually run on the layout, so please keep that in mind.[2]

CREATING LAYOUTS ON A COMPUTER

In the previous topic, we saw that even with only straight and curved tracks, you can create an enormous number of layouts. In fact, with 30 pieces, there are about 10 million possible combinations.

Counting all of these requires the help of a computer. To make them

---

[2]Some enthusiasts have even gone as far as using a 3D printer to create special crossing pieces, making it possible for trains to actually run on these intersecting layouts.

easier to handle, it is useful to represent each track type with a symbol. Let's assign S, R, and L to Straight, Right Curve, and Left Curve, respectively.

With this notation, the layouts shown in the figure below can be written as SRRLL, RRRSLLL, and RLRLLLLLL, starting from the • mark.[3]

We can make this even shorter by grouping consecutive symbols: for example, two consecutive R's become R2, and three consecutive L's become L3. With this convention, the same layouts can be represented as SR2L2, R3SL3, and RLRL6.

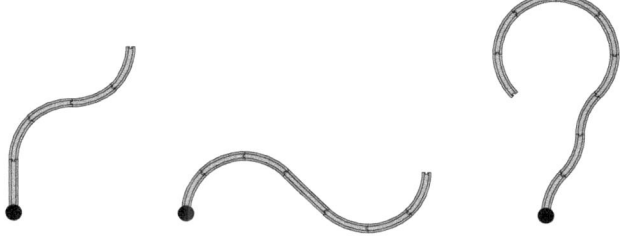

Figure 20.4 Track arrangements represented by SR2L2, R3SL3, and RLRL6.

With such simple rules, Plarail layouts can be represented using just symbols. In fact, I developed an application called the "Model Train Course Simulator," where you can input the letters S, R, and L, and the corresponding layout will appear on the screen. You don't even need to think too hard—just type in a random sequence of symbols and enjoy seeing the layout it creates.

This application is available on the web, so feel free to give it a try at the following address:

https://mitani.cs.tsukuba.ac.jp/ja/software/railway/index.html

---

[3]If only two types of tracks were enough, we could represent them with 0 and 1, which would feel much more "computer-like."

You can also scan the QR code below for quick access. It allows you to experiment with many different layouts before actually setting up the tracks in real life.

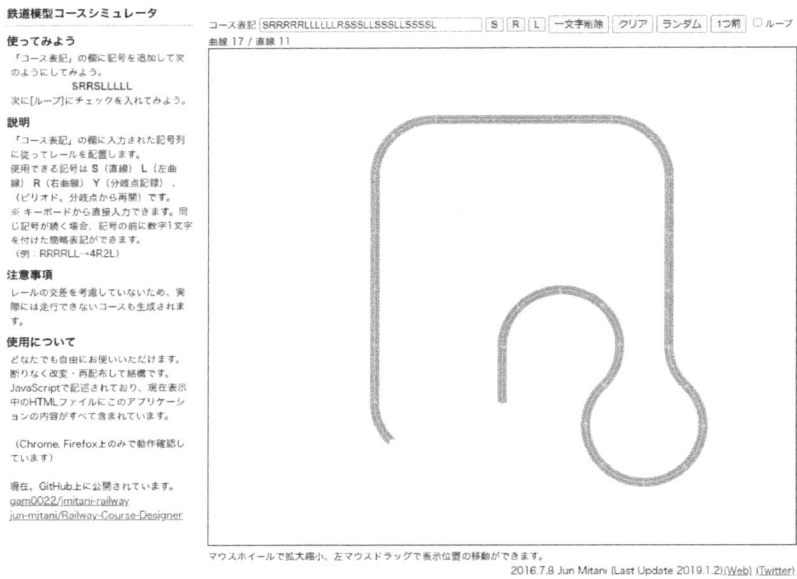

Figure 20.5 Screen of the model train course simulator.

The simulator also has an option called "Loop". When this option is turned on, it automatically generates closed courses using the method we discussed earlier.

In addition, I created another application called the "Geometric Pattern Course Random Generator." As shown in the figure below, this app continuously generates random geometric patterns. It feels a bit like playing with a kaleidoscope, so I highly recommend trying it out as well at the following address:

```
https://mitani.cs.tsukuba.ac.jp/ja/software/railway/index_
with_anime.html
```

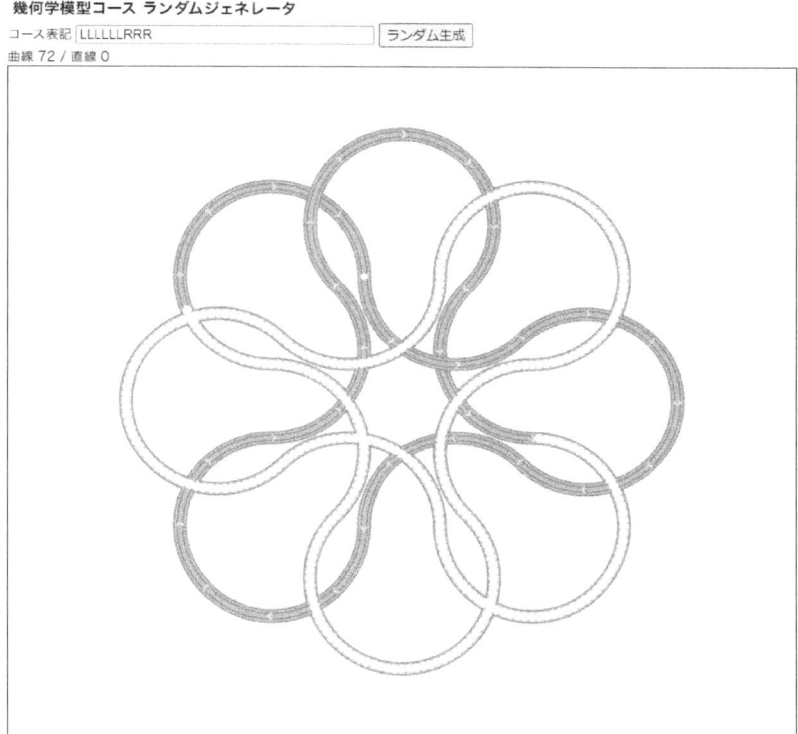

Figure 20.6 Screen of an app that generates geometric patterns drawable in a single stroke.

You can also scan the QR code below for quick access.

PUZZLE TO COMPLETE THE LAYOUT

Imagine you have a layout under construction, like the one shown in the figure below.

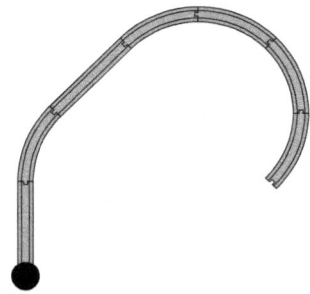

Figure 20.7 Layout under construction.

As we have seen, repeating a track sequence eventually creates a closed loop. But what if the challenge is different: "Complete the layout with as few additional pieces as possible"? This turns out to be quite a difficult problem.

An experienced builder might quickly figure out how to add the right tracks, but without that experience, you could spend a long time trying different combinations. Even when limited to just three types—right curves, left curves, and straight tracks—the number of possible combinations grows explosively with the number of pieces used, as we've already discussed.

I created a puzzle game based on this idea, where the goal is to complete an unfinished layout. It is a browser-based app that you can enjoy playing online. Please give it a try at the following URL:

`https://mitani.cs.tsukuba.ac.jp/ja/software/railway_puzzle/`

At the moment, there are 50 problems available. They get progressively more difficult toward the later stages. See how far you can go as a way to test your skills.

By practicing with this puzzle game, you'll likely find it much easier to complete the layouts your child leaves unfinished during play.

HOW WERE THE PUZZLE PROBLEMS CREATED?

I introduced the puzzle where the goal is to complete a layout. To be honest, even I, the developer, struggle with the later problems because they are so difficult. So how did I create them in the first place?

That's right—just as you might have guessed, both the problems and their solutions are generated by a computer. The program that finds the solutions is called a *solver*. Learning to program opens up possibilities like this, making it possible to create all sorts of fun and challenging things.

CHAPTER 21

Looking at Pi

Figure 21.1 What does this represent?

Can you figure out what the 3D shape in the figure above represents? Just by looking at its form, it may be difficult to guess. However, since the heading on this page contains the word "pi," those with a sharp intuition may already have figured out the answer.

PI

The ratio of a circle's circumference to its diameter, known as π, is an endless sequence of digits.

$$\pi = 3.14159265358979323846264 3 \ldots$$

For centuries, mathematicians have been fascinated by whether there might be hidden patterns in the digits of π. Along the way, people have tried many different methods of representing it in forms other than numbers, turning it into something more visual.

 DOI: 10.1201/9781003670261-21

The figure represents 1,000 digits after the decimal point as a 3D bar graph. The bars are arranged in groups of 50 digits each, moving from the front to the back. Although it may look like nothing more than a series of bars, seeing the digits of π transformed into a three-dimensional figure is quite intriguing.

FEYNMAN POINT

In this form, for example, it is easy to spot the so-called *Feynman Point*, where six consecutive 9's appear. The Feynman Point refers to the sequence of six 9's beginning at the 762nd decimal place of π. It is named after the American physicist Richard Feynman, who once joked during a lecture that he wanted to memorize π up to this point.

Here is a portion of the digits, from the 751st to the 770th place:

$$\ldots 51870721134999999837 \ldots$$

When represented as a bar graph, the cluster of 9's stands out clearly. In the next figure, this section is highlighted.

Figure 21.2 Diagram of Pi highlighting the Feynman point.

Incidentally, if digits were completely random, the probability of finding six identical digits in a row is only about 1 in 100,000—so encountering such a sequence so early in π is quite remarkable.

ENCODING PI WITH PLARAIL

In the previous topic, we explained how a Plarail layout can be represented as a sequence of three symbols: S, R, and L. By replacing these symbols with the numbers 0, 1, and 2 respectively, the layout RRSL shown in the figure below becomes 1102.

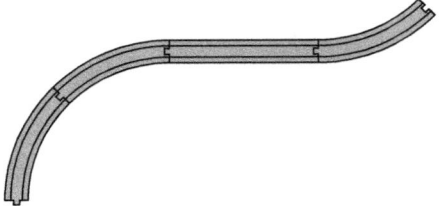

Figure 21.3 Layout represented by RRSL.

In everyday life, we use the decimal system, which represents numbers using combinations of ten digits from 0 to 9. However, it is also possible to use the *ternary system*, which represents numbers using only three digits: 0, 1, and 2.

For example, the number written as 2201 in ternary corresponds to 73 in decimal, as calculated below:

$$2 \times 3^3 + 2 \times 3^2 + 0 \times 3^1 + 1 \times 3^0 = 54 + 18 + 0 + 1 = 73.$$

The figure below shows a representation of the digits of π in ternary, specifically the sequence after the decimal point: 010211012222010211001120010201. By replacing each digit with the corresponding symbol—0 with S, 1 with R, and 2 with L—we convert it into the rail layout sequence: SRSLRRSRLLLLSRSLRRSSRRLSSRSSLSR.

In this way, we have expressed the number π through the shape of a Plarail layout.[1] This might be the first illustration to visualize the value of pi using toy railway tracks!

INFORMATION CONTENT OF PLARAIL

Encoding is the process of converting given information into a sequence of predefined elements, following specific rules. In our example, this means converting data into a sequence of rail pieces. Conversely, *decoding* is the process of recovering the original information—in this case, the digits of pi—from such a sequence.

Binary numbers, which use only two digits—0 and 1—are well known for storing *1 bit* of information per digit. In contrast, ternary numbers use three digits, and each digit can carry more information—about 1.6 bits. In general, when there are n possible values, the amount of information contained in a

---

[1] This corresponds to 0.1415926535 in decimal notation.

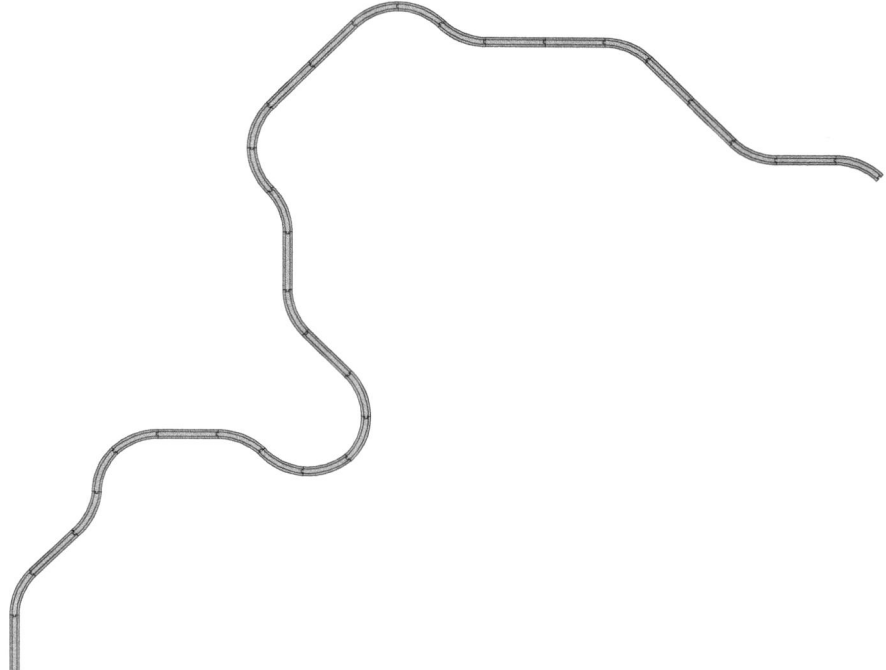

Figure 21.4 Layout encoding the digits of Pi.

single digit is given by $\log_2 n$ bits. So for a ternary digit, which has 3 possible values, the information content is

$$\log_2 3 = 1.5849\ldots.$$

This means that when we use only straight rails, right curves, and left curves to store data, each rail piece can represent approximately 1.6 bits of information.

CHAPTER 22

The Tower of Hanoi Algorithm

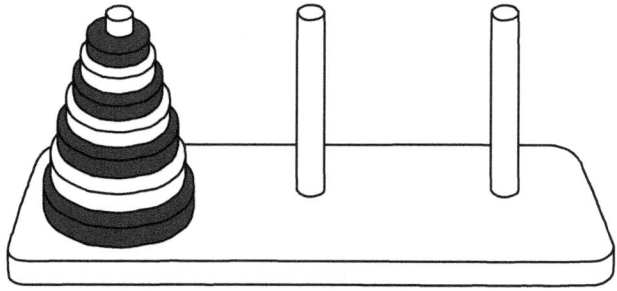

Figure 22.1   The tower of Hanoi puzzle.

There is a famous puzzle called "The Tower of Hanoi." In this puzzle, each disk has a single hole in the center and is stacked on a pole in descending order of size to form a tower. Once the tower is built on the leftmost pole, the goal is to move it to the rightmost pole. The puzzle comes with two rules: "only one disk can be moved at a time", and "a larger disk cannot be placed on top of a smaller one".

The rules are simple, but until you discover the solution, you'll find yourself twisting your head and trying different approaches. However, if the discs are colored as shown in the figure above (as is the case with the Hanoi Tower puzzle I have at home), this is a big clue. Once you see the pattern, you can solve the puzzle through a series of repeated, systematic moves. This is where the concept of *recursion* comes into play.

DOI: 10.1201/9781003670261-22

THE TOWER OF HANOI WITH 7 DISKS

Let's consider moving a 7-disk tower from the leftmost pole to the rightmost pole. We'll call the destination pole the "target pole," and the other pole (that is neither the starting nor the destination pole) the "auxiliary pole."

To move the largest disk at the bottom, we first need to move the 6-disk tower sitting on it to the auxiliary pole. Once that's done, we can move the largest disk to the target pole. Then, we move the 6-disk tower from the auxiliary pole to the target pole.

In summary, moving a 7-disk tower involves the following three steps:

1. Move the 6-disk tower to the auxiliary pole.

2. Move the largest disk to the target pole.

3. Move the 6-disk tower to the target pole.

Steps 1 and 3 both involve moving a 6-disk tower. But to do that, you first need to move a 5-disk tower. And to move a 5-disk tower, you need to move a 4-disk tower. And to move a 4-disk tower...

In this way, to move a tower with n disks, you need to move a tower with $n-1$ disks. And to move that, you need to move a tower with $n-2$ disks, and so on, all the way back to moving a single disk.

Now, let's take a closer look at how the movements work in the case of a 4-disk tower in the figure below.

As shown in the figure, the 1-disk tower, then the 2-disk tower, and then the 3-disk tower are moved in sequence. Only on the 8th move can the 4th (largest) disk finally be moved to the rightmost pole. After that, to move the 3-disk tower to the rightmost pole, you once again need to move the 2-disk tower, and before that, the 1-disk tower. The entire 4-disk tower can be moved in a minimum of 15 moves.

While the best way to understand this is by actually moving the pieces yourself, if you look closely at the diagram, you'll notice that the smallest disk is constantly on the move—hopping from one pole to another every other move.

ALGORITHM FOR SOLVING THE TOWER OF HANOI

The sequence of operations we discussed can be expressed in the following *pseudo-code*.

```
Function MoveTower(n, source, target, auxiliary)
    If n == 1 Then
        Move disk 1 from source to target
    Else
        // (1) Move the (n-1)-disk tower from source to auxiliary
        Call MoveTower(n - 1, source, auxiliary, target)

        // (2) Move the largest disk from source to target
        Move disk n from source to target

        // (3) Move the (n-1)-disk tower from auxiliary to target
        Call MoveTower(n - 1, auxiliary, target, source)
    End If

Start
    Call MoveTower(7, LeftPole, RightPole, MiddlePole)
End
```

The function MoveTower in this program moves a tower of height n from the source pole to the target pole. However, if you look inside the

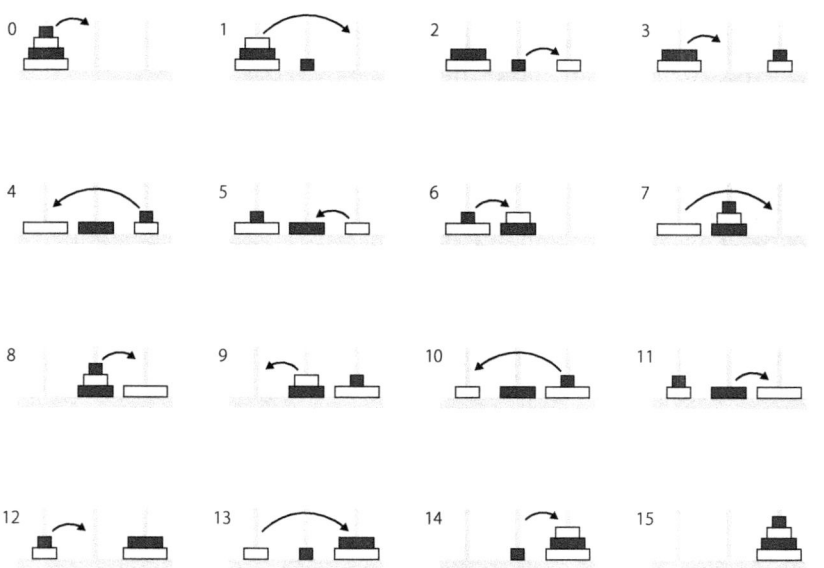

Figure 22.2 Steps for the tower of Hanoi with 4 disks (the number n indicates the state after the n-th move).

function, you'll see that it calls itself twice. This kind of function, where the function calls itself within its own definition, is called a *recursive function*.

Understanding code that uses recursion can be a bit tricky at first, but if you follow the flow step by step, you'll see that it faithfully replicates the exact process needed to solve the Tower of Hanoi. Even though solving the Tower of Hanoi involves a complex series of repeated steps, it can be accomplished with such a simple algorithm.

Earlier, I mentioned that the colors of the disks can serve as a hint for solving the puzzle. As you can observe from the diagram showing each step, at no point do disks of the same color ever end up stacked on top of each other. In other words, a white disk is never placed on another white disk, and a black disk is never placed on another black disk.

Once you understand this rule, it becomes clear that there is only one possible destination for each disk at every step. This means you can successfully move the entire tower without any hesitation or confusion.

MINIMUM NUMBER OF MOVES REQUIRED TO SOLVE THE TOWER OF HANOI

It is known that for a Tower of Hanoi puzzle with n disks, the minimum number of moves required is $2^n - 1$.[1]

This means that for 3, 4, 5, 6, and 7 disks, the required number of moves is 7, 15, 31, 63, and 127, respectively. With each additional disk, the number of moves nearly doubles.

For example:

- 10 disks require 1,023 moves

- 20 disks require 1,048,575 moves

- 30 disks require 1,073,741,823 moves

Even if you could make one move per second, it would take approximately 34 years to complete the Tower of Hanoi with 30 disks!

---

[1]Numbers of the form $2^n - 1$ are called *Mersenne numbers*.

CHAPTER 23

Folding Paper Along Curves Is Fun

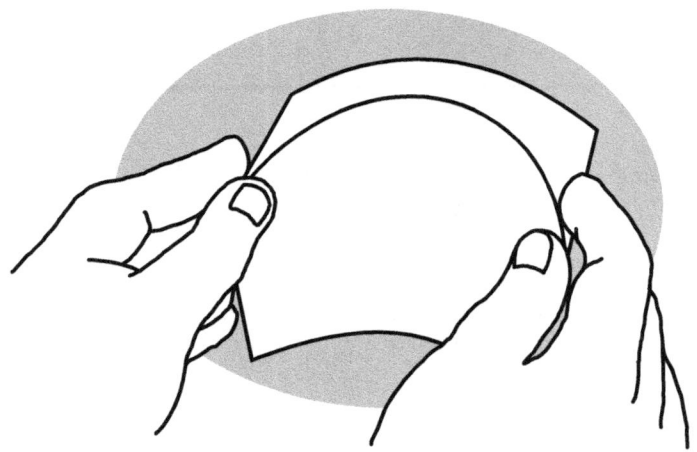

Figure 23.1 Trying to fold paper along a curve.

If you're handed a single sheet of paper and told to fold it however you like, most people will probably fold it by matching corner to corner or edge to edge, creating a flat, crisp fold. Someone with a slightly more unconventional approach might fold it somewhere at random without aligning any edges—but even then, they'll likely fold it flat. And when a fold is flat, the crease that forms is straight—a line.

DOI: 10.1201/9781003670261-23

But paper can also be folded along a curve. Try drawing a random curved line on the paper and folding along it. You'll discover that this creates shapes you don't normally see in everyday folding.

FOLDING ALONG A CURVE

When paper is folded along a curve, smooth curved surfaces appear on both sides of the crease. The result is not flat—it takes on a three-dimensional form. The shape that emerges depends on the angle at which you fold along the crease.

In the photos below, all three forms were made by folding the same curve, yet each one gives a very different impression.

Figure 23.2 Folding the same curve can create different shapes.

Curved surfaces that can be formed by bending paper—without stretching or compressing it—can be unfolded back into a flat sheet. These are called *developable surfaces*. They are a type of *ruled surface*, as introduced in the topic "Chapter 18: The Mysterious Surface Made of Cotton Swabs: A Hyperboloid."

The figure below shows one of my origami works titled "Lysichiton," made by folding a single curve on an A4-size sheet of paper.[1]

Take a sheet of copy paper and draw a curve similar to the one on the left side of the figure. Once you've creased along the curve, pinch the top-right and bottom-right corners of the paper with your right and left hands, respectively. Then, by gently twisting the entire sheet, the

---

[1]This origami piece is included in the book: Jun Mitani, Curved-Folding Origami Design, A K Peters/CRC Press, 2019.

final shape will emerge. It's not easy to explain in words, so give it a try yourself. All you need is a sheet of copy paper—so go ahead and see what you can create!

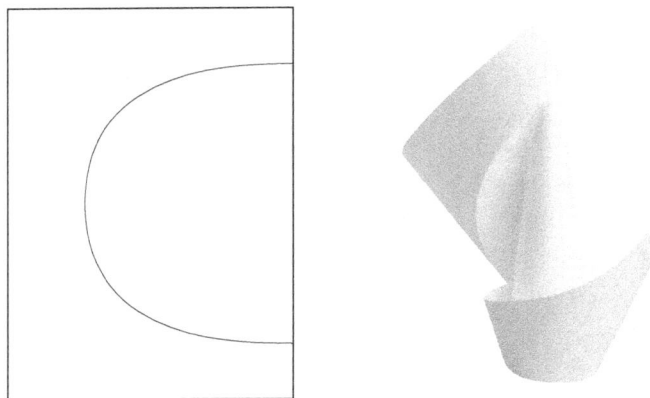

Figure 23.3 An artwork of a "Lysichiton" that can be made using A4-size paper.

FOLDING INTO GEOMETRIC SHAPES

It turns out that even if you draw a curve freehand, folding a sheet of paper along it can create interesting shapes. However, this method cannot produce exactly the shape you want. By determining the curve precisely through calculations—based on the geometric restrictions of shapes that can be made from a single sheet of paper—you can create the intended shape with accuracy.

This might sound a little complicated, but if we limit ourselves to making *axisymmetric solids* and fold the paper so that the pleats extend outward, then with some simple calculations we can design fold lines along curves to form such shapes. I will leave the detailed explanation of this approach to another book,[2] and instead introduce two origami works created using curves designed in this way.

The figure below shows an example where a long, narrow rectangle is transformed into a sphere. To assemble it, paste together the ● and ●, as well as the ◯ and ◯, to make a cylindrical form, and then finish it up. Solid lines represent mountain folds, while dashed lines represent valley folds.

[2] Jun Mitani, 3D Origami Art, A K Peters/CRC Press, June 2016.

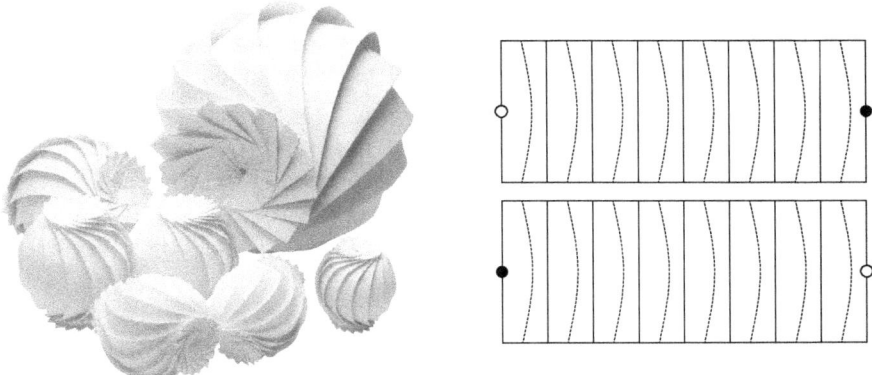

Figure 23.4 Spherical origami.

The figure below shows a shape made to look like whipped cream.[3] Both of these are forms that I am quite fond of.

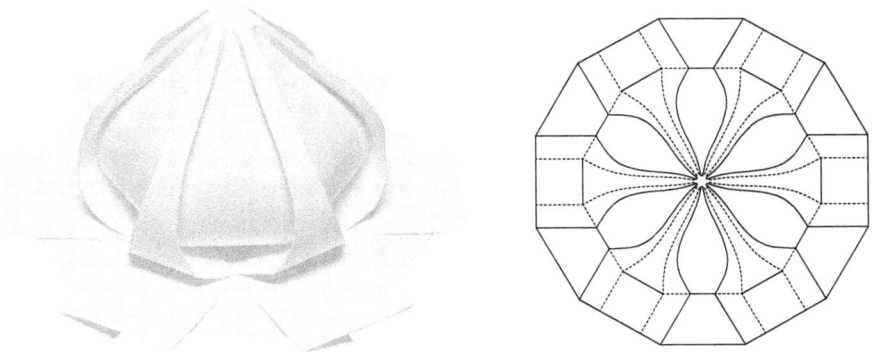

Figure 23.5 Origami shaped like whipped cream.

FOLDING A PAPER STRIP

As we saw earlier, by freely drawing a curved fold line on a sheet of paper, you can create many different shapes. However, if the line is too wavy, it may become difficult to fold neatly along it.

---

[3]Depending on the viewer, however, they might appear to resemble other things—such as a peach or a *gibōshi* (a type of ornamental finial).

This may still be hard to picture just from words, so please try it out yourself. Draw a squiggly line on a sheet of paper and see whether you can actually fold along it. You will notice that the kinds of curves that fold cleanly are surprisingly limited.

When it becomes difficult to make a neat fold, one useful trick is to cut the paper into a long, narrow strip around the fold line, as shown in the figure below. Interestingly, it is known that in this form, any curve can be folded successfully.[4]

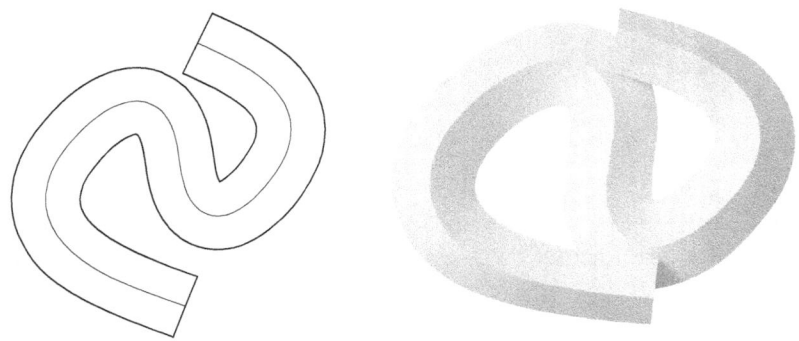

Figure 23.6 Even very wavy curves can be Folded if the width of the paper is made narrow.

As shown in the next figure, if you increase the sharpness of the fold along the crease, the original curve will bend more tightly. In other words, the greater the folding angle, the larger the curvature of the crease becomes.

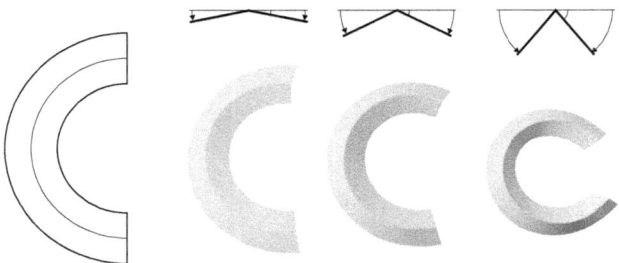

Figure 23.7 Differences between folding angles and the resulting shapes.

---

[4]Reference: Dmitry Fuchs and Serge Tabachnikov. More on Paperfolding. The American Mathematical Monthly 106:1 (1999), 27-35.

By making good use of what we have discussed so far, you can even create a treble clef from a single continuous strip of paper, as shown in the figure below. At first glance, the treble clef seems impossible to form from just one piece of paper, since the lines appear to cross each other—but in fact, it can be done.

Figure 23.8 Treble Clef made with a long, narrow strip of paper.

Cut the paper into a long, narrow strip and place a crease down the center. By adjusting the folding angle of this crease, you can fine-tune the way the strip curves. This approach is quite different from ordinary origami, but it opens up a wide range of possibilities—and that freedom makes it all the more enjoyable.

CHAPTER 24

Folding Paper to Reach the Moon

Figure 24.1 Can we fold a sheet of paper until its thickness reaches the Moon?

When you keep folding a sheet of paper in half again and again, its thickness doubles each time. Eventually, it would surpass the height of Mt. Fuji (the tallest mountain in Japan at 3,776 meters), and by the

DOI: 10.1201/9781003670261-24

42nd fold, it would even reach all the way to the Moon. You may have heard this story at least once before.

LET'S ACTUALLY DO THE CALCULATION

Let's check whether such a thing is really possible by doing the math, assuming the thickness of the paper is 0.1 millimeters. Each time you fold it, the total thickness doubles. Below, we'll list the number of folds and the thickness (or height) at that point.

Number of times folded	Height
1	0.2mm
2	0.4mm
3	0.8mm
4	1.6mm
⋮	⋮
10	10cm
⋮	⋮
23	839m
⋮	⋮
25	3,355m
26	6,710m
⋮	⋮
42	439,804km

Skytree
634m

Mt. Fuji
3,776m

◀ Taller than Skytree (the tallest structure in Japan)

◀ Almost as tall as Mt. Fuji

◀ Surpassing Mt. Fuji

◀ Exceeding the distance to the moon (384,400 km)

Indeed, by the 42nd fold, the thickness exceeds the distance to the Moon. This kind of growth, where the amount doubles each time, is called *exponential growth*. At first, the increase may seem small, but before you know it, the numbers become enormous. The power of doubling is truly astonishing.

WHAT HAPPENS TO THE SIZE OF THE PAPER?

In the earlier discussion we only focused on the thickness of the paper and didn't think about its size. Since we are folding it in half again and again, the area of the paper is naturally reduced to half each time.

Let's calculate how the size of the paper changes if we start with a single sheet of copy paper. Suppose we fold it alternately horizontally and vertically. We'll look at how the length of the longer side changes as we go.

An A4 sheet of copy paper has an aspect ratio of $1 : \sqrt{2}$ (known as the *silver ratio*). A special property of this ratio is that even after folding the long side in half, the proportion is preserved. After one fold, the area becomes 1/2 of the original, and the long side becomes $1/\sqrt{2}$ of its original length. After two folds, the area becomes 1/4, and the long side becomes half of the original. Since the long side of an A4 sheet is about 30 cm, after one fold the long side is about 21 cm, and after two folds it becomes about 15 cm.

Below, we will list the number of folds and the corresponding paper size (the length of the long side).

Number of times folded	Paper size (length of long side)	
1	21cm	◀ A5 size
2	15cm	◀ Postcard size
3	10.5cm	
4	7.5cm	
⋮	⋮	
8	18.5mm	◀ About the size of a penny
⋮	⋮	
20	0.3mm	◀ The thickness of mechanical pencil leads
⋮	⋮	
28	20 μm	◀ Thickness of paper fibers
⋮	⋮	
42	0.1 μm	◀ Virus size

As shown in the table, by the 28th fold the size of the paper would become thinner than the width of a single paper fiber. By the 42nd fold—the one that supposedly reaches the Moon—the paper would shrink to something on the scale of a virus. Clearly, then, the idea of folding a sheet of paper until it reaches the Moon is not something that can actually be done. You probably suspected that already.

If you try it with real copy paper, you'll find that the practical limit is about seven folds. This happens because, as shown in the illustration below, the folded edges start to curve and round off, and most of the paper ends up being taken up in those folded sections.

Figure 24.2 We need to take the folded part into account.

DRAGON CURVE

If you prepare a long strip of paper, fold it in half again and again, and then open each crease at a right angle as shown in the figure below, an interesting shape appears. This shape is known as the *dragon curve*.

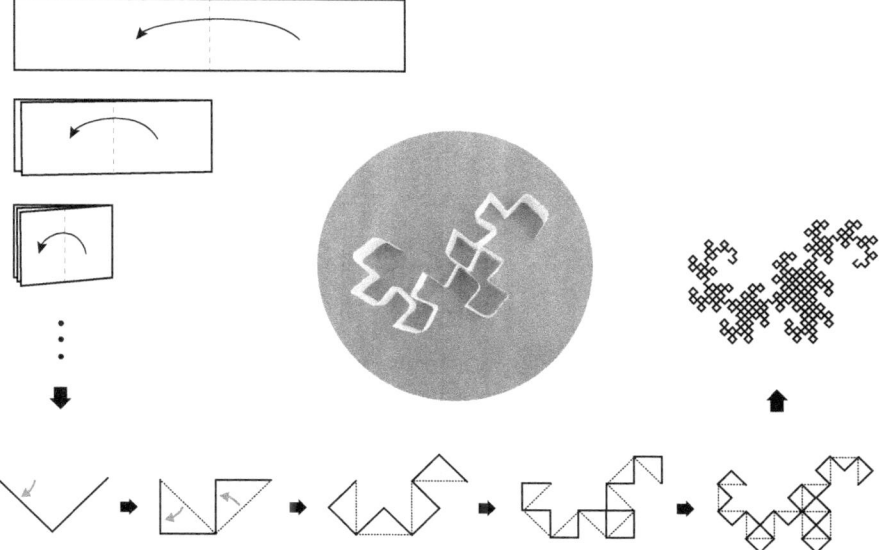

Figure 24.3 Creating a Dragon Curve.

The dragon curve is a type of *fractal figure* with self-similarity, meaning that its parts and its whole share the same structure. It has the remarkable property that it never intersects with any part of itself. Another fascinating feature is that if you take copies of it and rotate them by 90°, 180°, or 270°, they can tile the entire plane without leaving any gaps. You can observe this simply by folding and unfolding paper, so I encourage you to give it a try.

The name "dragon curve" has an interesting origin. In 1966, John Heighway, a physicist working at NASA, noticed the fascinating pattern that appeared when he folded and unfolded a one-dollar bill. One of his colleagues then gave this shape the name dragon curve.

AREA OF A-SERIES PAPER

Labels such as A3, A4, and A5 are commonly used to indicate paper sizes, like "A4 copy paper." The size of A4 paper is $297\,\text{mm} \times 210\,\text{mm}$, and this is defined by an international standard.

A key feature of this standard is that the aspect ratio is $1 : \sqrt{2}$. If you cut the sheet in half along its longer side, the aspect ratio remains the same while the area becomes half. As shown in the figure below, each time the number in the paper size label increases by one, the side lengths are reduced by a factor of $1/\sqrt{2}$, and the area is reduced by half.

The area of an A0 sheet ($841\,\text{mm} \times 1189\,\text{mm}$) is exactly $1\,\text{m}^2$.[1] That's a pretty neat system, isn't it?

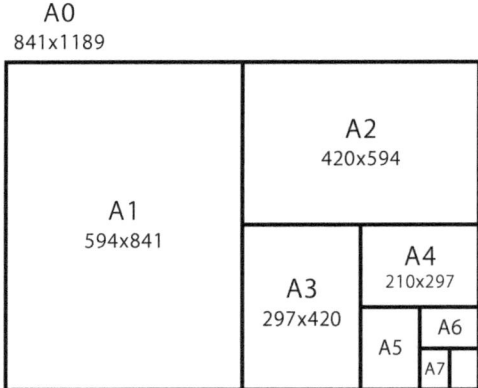

Figure 24.4 A-series paper sizes.

---

[1]If you actually do the calculation, you'll find that the area is $0.999949\,\text{m}^2$. Since the dimensions are specified in millimeters, the ratio is not exactly $1 : \sqrt{2}$ in the strictest sense.

CHAPTER 25

The Wonder of Repeated Folding

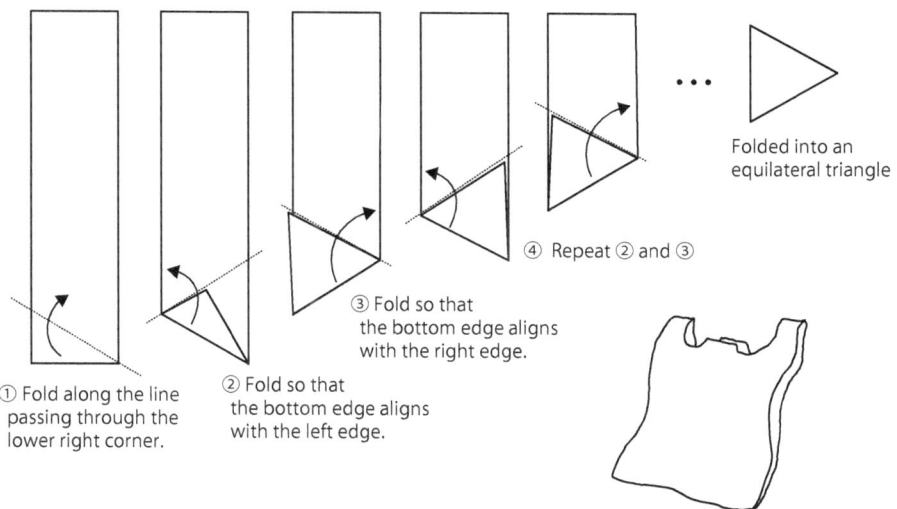

Figure 25.1 How to fold a plastic shopping Bag into a triangle.

One method of folding a plastic shopping bag is to fold it into a triangle. You first fold the bag into a long, narrow strip, and then, as shown in the diagram above, begin folding it into triangles from one end.

What's amazing about this folding method is that no matter how randomly you start, it will almost always end up as a nearly perfect equilateral triangle.

DOI: 10.1201/9781003670261-25

WHY IT APPROACHES AN EQUILATERAL TRIANGLE

When you fold a strip in the way shown at the beginning, even if you start with a slightly distorted triangle, the shape gradually approaches an equilateral triangle. It's quite mysterious.

Try it yourself using a piece of copier paper. Cut out a long, narrow rectangle and start folding it into triangles. You'll find—surprisingly—that it really does form an equilateral triangle. So, how does this work?

For the triangle to be equilateral, the angle between the edge of the strip and the fold must be exactly 60°. In the initial diagram, the first fold is made more or less arbitrarily, so let's assume there is an error of α.

Then, the angle labeled ① in the diagram below would be:

$$180° - (60° + \alpha) = 120° - \alpha.$$

In the next step, you fold to bisect this angle, so the resulting angle becomes:

$$60° - \frac{\alpha}{2}.$$

The error is now halved.

Next, angle ② becomes:

$$180° - \left(60° - \frac{\alpha}{2}\right) = 120° + \frac{\alpha}{2}.$$

When this is bisected in the following fold, the resulting angle becomes:

$$60° + \frac{\alpha}{4}.$$

Again, the error is halved.

After three folds, the error is reduced to one-eighth, and after four repetitions, it becomes one-sixteenth. In other words, even if the initial error were as large as 15° (which is quite a significant deviation), after four folds the error would shrink to less than 1°.

In this way, you can eventually fold the strip into a neat equilateral triangle. This process is called *asymptotic* convergence to an equilateral triangle. It's quite enjoyable to realize that, by simply repeating the same folding motion, anyone can end up with an almost perfectly equilateral triangle.

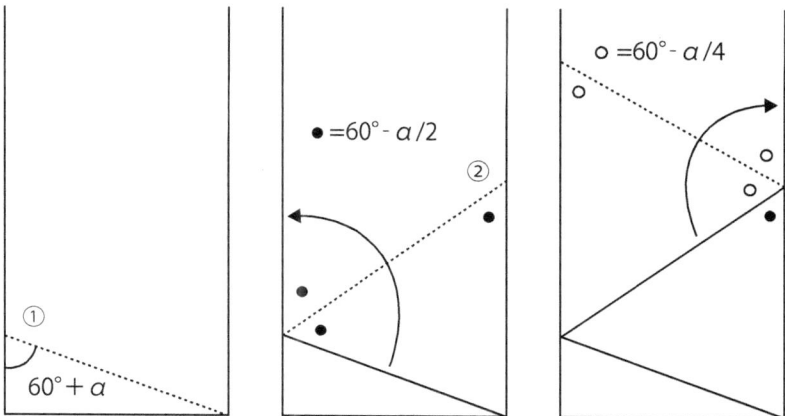

Figure 25.2 Changes in angles through folding.

DIVIDE INTO THREE EQUAL PARTS BY FOLDING

There are other interesting results you can get by repeating folding operations.

For example, if you follow the method shown in the next diagram, the crease will gradually approach the point that is one-third of the way across the paper's width.

This is a neat trick for dividing a sheet into thirds. Each time you fold, the initial error is halved. So after just a few repetitions, you'll end up folding at a point so close to exactly one-third that it's accurate enough for practical purposes. Definitely give it a try!

This kind of repeated folding method—for dividing paper into n equal parts or for locating a position like n/m—is known as the *Fujimoto Asymptotic Method*, named after its pioneer, Shuzo Fujimoto.

FOLDING THAT REVEALS A QUADRATIC CURVE

Creases made by folding a sheet of paper flat are always straight lines, but with those straight creases, you can actually create curves such as parabolas or ellipses. Here's how you can draw a parabola using folds:

1. Place a point p somewhere on the paper.

2. Choose a point q on the edge of the paper, and fold the paper so that point q lands on point p.

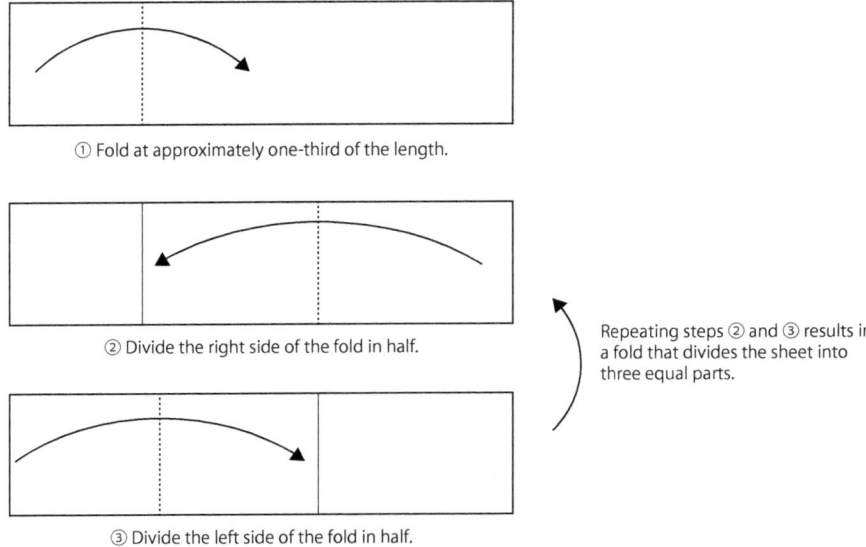

① Fold at approximately one-third of the length.

② Divide the right side of the fold in half.

③ Divide the left side of the fold in half.

Repeating steps ② and ③ results in a fold that divides the sheet into three equal parts.

Figure 25.3 How to fold into three Equal Parts.

3. Repeat step 2 while moving point q along the edge of the paper.

In this way, the crease becomes a tangent line to a parabola whose *focus* is the point p and whose *directrix* is the edge of the paper. As shown in the figure, when all these lines are taken together, the shape of the parabola emerges. Such a collection of lines is called the *envelope* of the parabola.

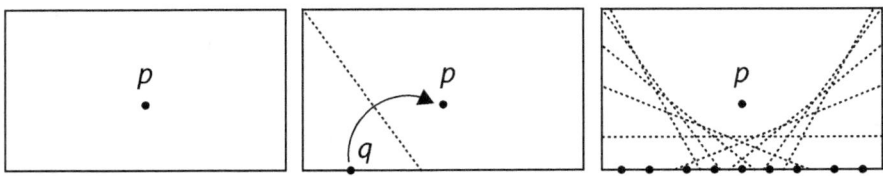

Figure 25.4 Folding the envelope of a parabola.

When we think of origami, we usually imagine using a square sheet of paper. But if you use a circular sheet instead, you can create the envelope of an ellipse in exactly the same way as we did earlier for the parabola. The only difference is the shape of the paper—the steps themselves are completely the same.

1. Place a point p somewhere on the paper.

2. Choose a point q on the edge of the paper, and fold the paper so that point q lands on point p.

3. Repeat step 2 while moving point q along the edge of the paper.

The creases form the envelope of an ellipse whose *foci* are point p and the center of the original circle.

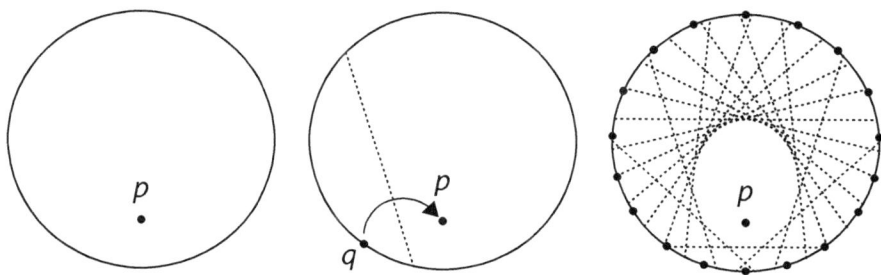

Figure 25.5 Folding the envelope of an ellipse.

As we saw in the topic "Chapter 4: Conic Sections Created by Light," both the parabola and the ellipse are conic sections. By changing the shape of the paper, we were able to fold envelopes of these two curves. But what about the third conic section, the *hyperbola*—how can we fold it?

For the ellipse, we folded points on the circumference of a circle so that they matched a point inside the circle. This time, if we instead fold points on the circumference so that they land on a point *outside* the circle, we obtain the envelope of a hyperbola.[1]

This may require using semi-transparent tracing paper, or alternatively, you can cut a circular hole in the paper to make the folding process possible.

---

[1]Reference: Yoshiyuki Yokota, Origami and Mathematics Education, Bulletin of Teacher Training Course, Tokyo Metropolitan University, (2), pp.181–193, 2018. (in Japanese)

CHAPTER 26

Modular Origami Balls and Polyhedral Duals

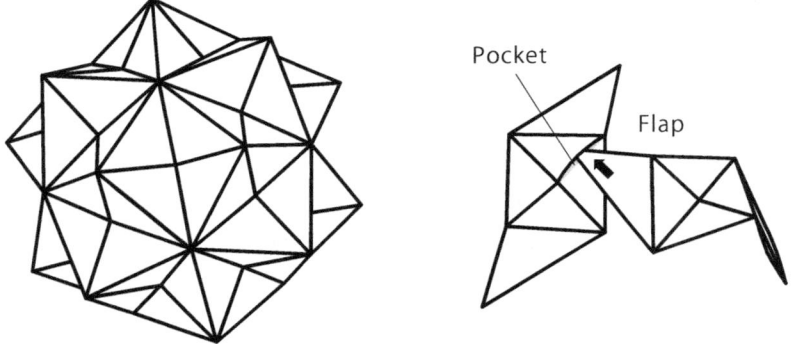

Figure 26.1 Modular origami ball and *sonobe Units.*

Origami works in which identical folded units are assembled into a solid are called *modular origami*. A well-known example is the *origami ball* constructed from pieces called the *Sonobe Unit.*[1]

In addition to the Sonobe Unit, many other types of units have been invented, leading to a wide variety of possible assembled forms. The process of putting the units together is enjoyable, but it requires patience to prepare accurately folded several dozen units in advance (Above all,

---

[1]In Japan, origami balls are called "Kusudama." The Sonobe Unit was first used by Mitsunobu Sonobe in his work called "Color Box," which was originally assembled from six units. Later, larger constructions using 12 units, 30 units, and more were developed.

 DOI: 10.1201/9781003670261-26

the trickiest part is carefully fitting in the very last unit to complete the structure.)

So, how many units are actually needed to complete a modular origami piece? Of course, it depends on the solid you are making, but to put it rather bluntly, in most cases the answer is "30 units."

SOLIDS BUILT FROM ORIGAMI UNITS

Most modular origami models are based on regular polyhedra, since the units are identical and fit together with a high degree of symmetry. As introduced in "Shapes Made with Triangles," there are only five regular polyhedra: the tetrahedron, cube, octahedron, dodecahedron, and icosahedron. Among them, the icosahedron is the closest in shape to a sphere, followed by the dodecahedron.

In modular origami, the units are often arranged "along the edges" of a polyhedron. Interestingly, both the dodecahedron and the icosahedron have exactly 30 edges. This is why so many modular origami constructions require exactly 30 units—hence the common rule of thumb to prepare 30 units in advance (yes, quite a lot of folding!).

The following table, also shown in "Chapter 13: Shapes Made with Triangles," reviews the number of vertices, edges, and faces of each regular polyhedron.

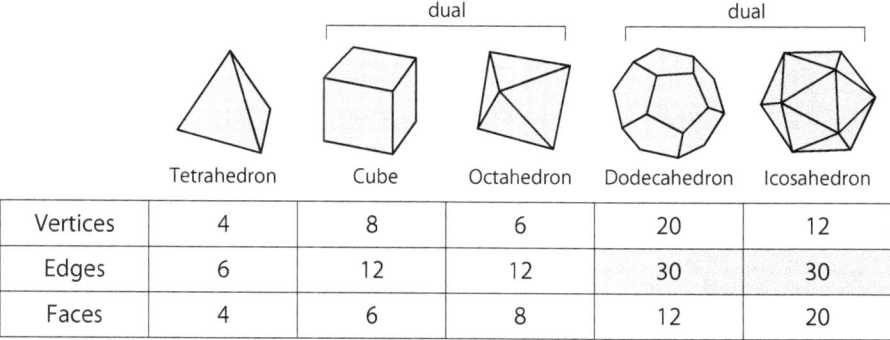

	Tetrahedron	Cube	Octahedron	Dodecahedron	Icosahedron
Vertices	4	8	6	20	12
Edges	6	12	12	30	30
Faces	4	6	8	12	20

Figure 26.2 The number of vertices, edges, and faces of regular polyhedra.

The "kusudama" shown at the beginning is based on the icosahedron, with each unit positioned along one of its edges. For this reason, it also requires 30 units to complete.

DUALS OF POLYHEDRA

Looking at the table again, we can notice an interesting relationship. As mentioned earlier, the dodecahedron and the icosahedron have the same number of *edges*, but if we look instead at the *faces* and *vertices*, we find that they are exactly swapped.

Similarly, if we compare the cube and the octahedron, they also share the same number of edges, while their numbers of faces and vertices are exchanged.

These pairs—the dodecahedron and icosahedron, the cube and octahedron—are said to be in a relationship called *duality*. To form the dual of a polyhedron A, we place a new vertex at the center of each face of A, and then connect the new vertices corresponding to adjacent faces. The resulting polyhedron B is the dual of A.

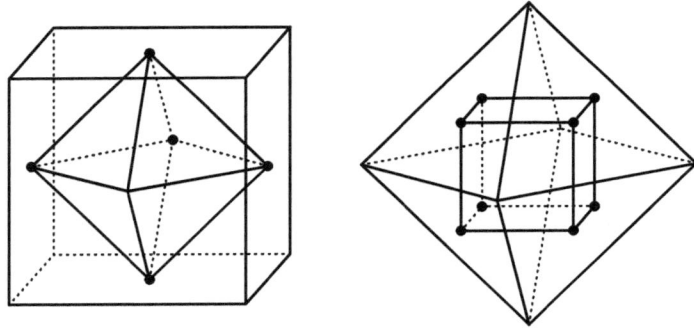

Figure 26.3 Duality between the cube and the octahedron.

The tetrahedron is the only regular polyhedron without a distinct dual partner—it is dual to itself. If you try replacing each face of a tetrahedron with a vertex and then construct the corresponding solid, you end up with another tetrahedron. A solid with this property is called *self-dual.*

Knowing these properties of regular polyhedra can be very helpful when you take on the challenge of designing new modular origami models.

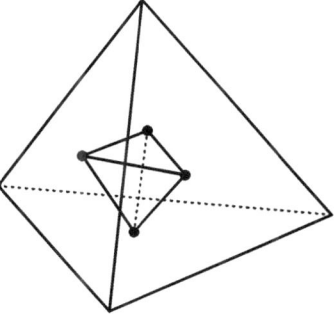

Figure 26.4 The tetrahedron is a self-dual solid.

SEMIREGULAR POLYHEDRA (ARCHIMEDEAN SOLIDS)

There are only five regular polyhedra. Each of them is made entirely of a single type of regular polygon. However, if we allow different kinds of regular polygons to be combined—under the condition that the arrangement of faces around each vertex is the same—then we can create 13 additional polyhedra. These are called *semiregular polyhedra*, or *Archimedean solids*.

One well-known example is the *truncated icosahedron*, which is formed by cutting off the vertices of an icosahedron. Its faces consist of regular pentagons and hexagons, and it is the shape of a classic soccer ball.

Such polyhedra are also sometimes used as the underlying structures for modular origami. Depending on which polyhedron is chosen, the number of required units will vary. Thinking about how many units are needed for each case can be quite fun.

CHAPTER 27

Pop-Up Figures

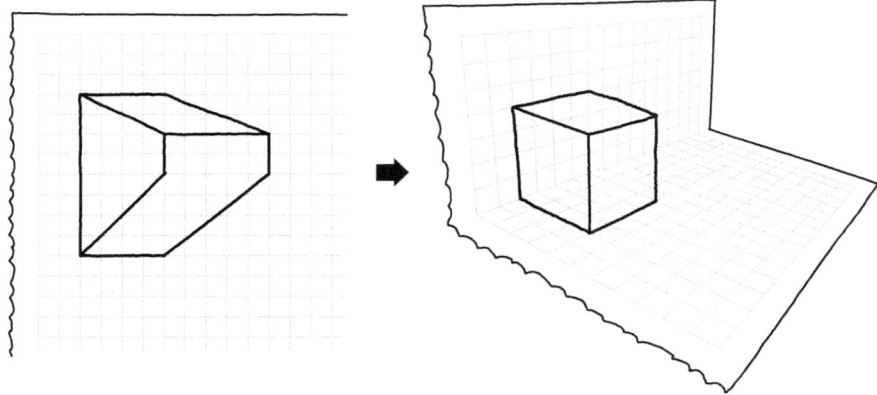

Figure 27.1 When graph paper with a figure (left) is folded upright along a horizontal line. . . .

On the left side of the figure above, you see a curious shape drawn on graph paper. At first glance, it's hard to tell what it represents. But when you fold the paper upright along a horizontal line—suddenly, something strange happens. It looks as if a cube is floating in the air.

This is one example of an optical illusion figure, where a three-dimensional shape appears when viewed from a particular position.

DOI: 10.1201/9781003670261-27

SHAPES THAT POP OUT WHEN THE PAPER IS FOLDED UPRIGHT

The cube illusion can be easily reproduced by anyone: simply copy the figure shown at the beginning onto graph paper, carefully following the grid. Give it a try!

Once you've drawn the figure and folded the paper along the horizontal line, step back a little and look at it with one eye closed—or view it through your smartphone camera. The effect of the three-dimensional cube floating in space becomes even stronger. Fascinating, isn't it?

So then, how can we represent more complex solids instead of just a cube? How can we design the flat figure, like the one shown on the left side of the opening diagram, before folding?

While you could try to work out some rules and draw it systematically, the following simple procedure works very well:

1. Take a photo of the folded graph paper.

2. Draw the desired figure on top of the photographed graph paper (you can either print the photo and draw directly on it, or sketch digitally on the computer screen).

3. For each straight line of the drawn figure, check which row and column of the grid its endpoints fall on, and then copy the line onto the original unfolded graph paper at those positions.

4. Look at the paper from the same angle as when you took the original photo.

Using this approach, I was able to create figures that appear to pop out as cubes and rectangular prisms, as shown in the next diagram. Give it a try yourself—you'll find it both fun and surprising!

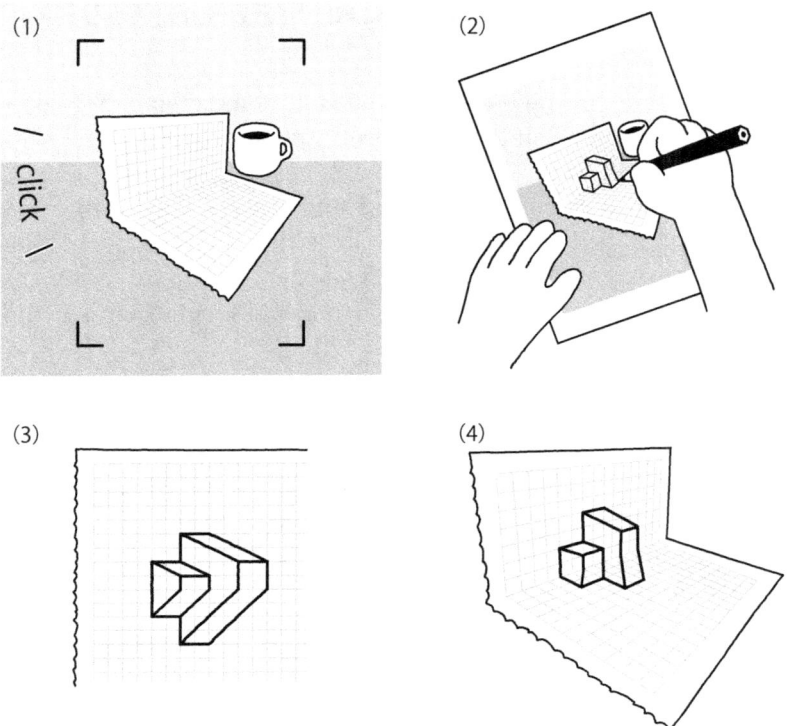

Figure 27.2 How to draw pop-up optical illusions.

POP-UP CARD

The figures we looked at only appeared to rise out of the paper, but in reality they were just flat drawings on the surface.

In contrast, there are also many *pop-up cards* sold commercially where real three-dimensional structures actually lift up from the paper. When you open a folded card, charming shapes designed for birthdays, Christmas, and other occasions spring to life in 3D. There are also many pop-up books that make full use of this mechanism, delighting readers with pages where entire scenes literally pop out as you turn them.

Here, let's consider making a simple pop-up card like the one shown in the next figure. The restriction is that we use only a single sheet of paper, adding just straight cuts and folds so that, when the card is opened to a right angle, a shape rises up.

This type of card can be folded flat in half, and when you open it, the three-dimensional form stands up.[1]

Figure 27.3 Pop-up card template made with a single sheet and simple cuts.

As the simplest example, let's look at the design where a rectangular prism pops up, as shown in the figure below (since it is made from a single sheet of paper, the side faces are not present).

---

[1]Masahiro Chatani's "Origamic Architecture," in which buildings rise up from pop-up cards, is world-famous.

From the side view, as shown on the right, you can see how the mechanism works: the square flattens into the sheet by moving through the shape of a parallelogram, just like a *parallel-link mechanism*.

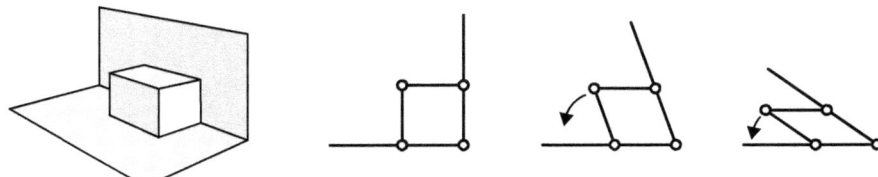

Figure 27.4 Side view of a pop-up card with a rectangular prism standing up.

The shape shown below, made by taking only the top and front faces of a solid built from stacked cubes, can also be created from a single sheet of paper with just a few cuts. Like the earlier example, it can be folded completely flat.

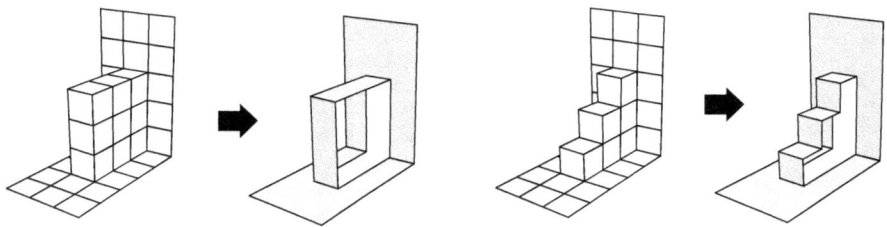

Figure 27.5 Creating a pop-up card shape from the top and front faces of stacked cubes.

Once you understand the principle, you can combine this mechanism in different ways to create a wide variety of shapes.

HOW TO MAKE TEMPLATES

How can we design a template for these pop-up cards? Here, let's go through the concrete steps of drawing a template, with reference to the figure below.

1. Prepare the base layout for a rectangular sheet that will be folded in half. At this stage, it only needs a single horizontal valley fold line in the center.

2. Using that fold as the baseline, draw the shape you want to rise up.

3. Suppose the shape is meant to stand forward from the backside by a distance a. Shift the entire shape downward by a.

4. To connect the shape to the backside when it stands up, place a rectangle of height a above the top edge of the shape.

5. Finally, convert the horizontal lines into fold lines. The template is now complete.

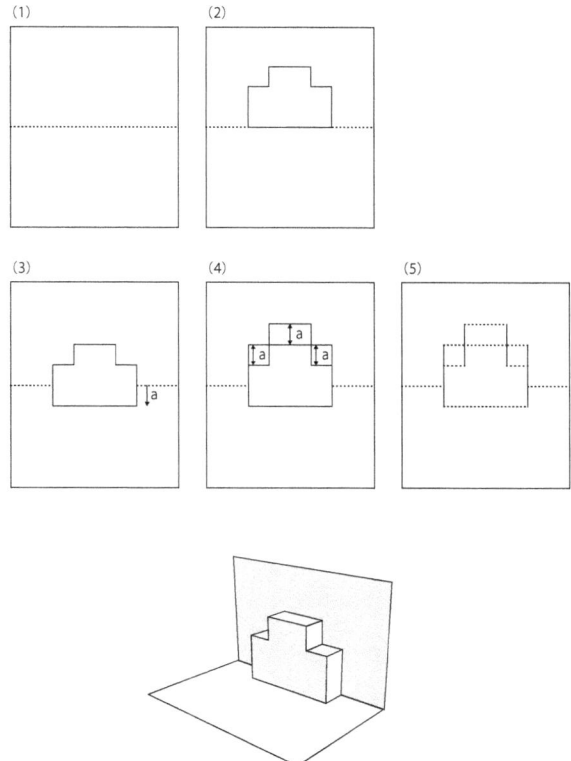

Figure 27.6 How to make pop-up cards(1).

By repeating the same kind of steps, you can add more shapes. Continuing from the earlier procedure, try the following:

6. Draw the new shape you want to add, aligning their bottom edges to the figure you just created.

7. Depending on how far forward you want this new shape to stand, shift it downward by a distance b.

8. To connect this new shape to the previous one, place a rectangle of height b above its top edge.

9. Finally, change the horizontal lines into fold lines, and the design is complete.

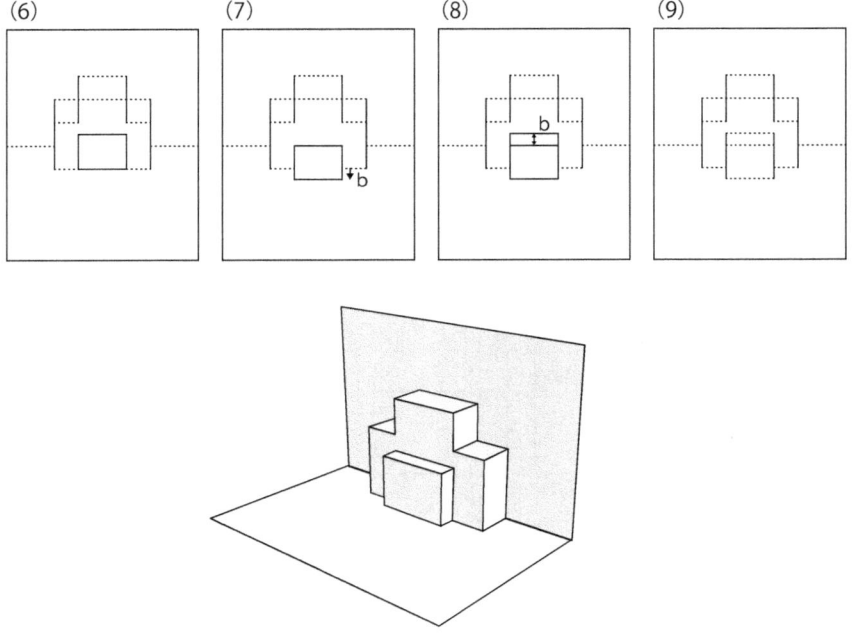

Figure 27.7 How to make pop-up cards(2).

Once you understand these steps for making pop-up cards, try creating an original design of your own!

APPS FOR EASY DESIGN

The best way to really understand how the mechanism works is to draw the templates by hand. However, if you'd like to make them more easily on a computer, I recommend the web app I created called "Pop Up Block Designer." It lets you design pop-up cards by stacking cube-like blocks. If you're interested, please give it a try!

https://mitani.cs.tsukuba.ac.jp/ja/software/popup_card/

CHAPTER 28

The Shape of a Peel Spun from an Apple

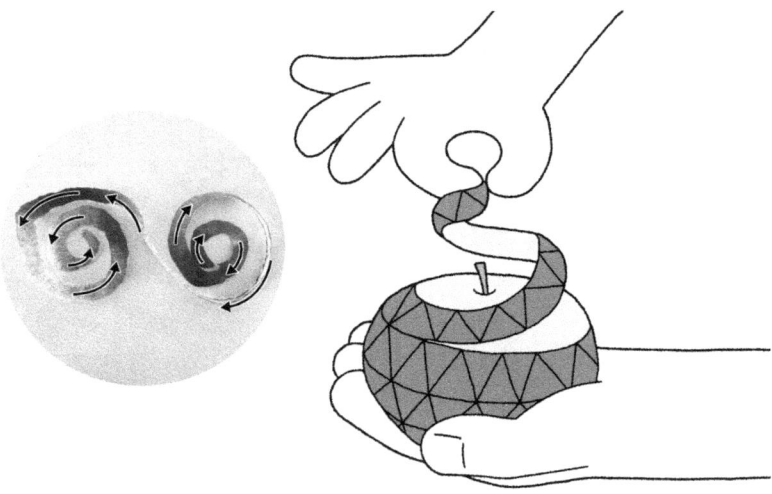

Figure 28.1 The look of a peel twirled off an apple.

When you peel an apple while twirling it, the peel becomes one long, continuous strip. If you lay it flat, it takes the shape of two spirals connected in opposite directions. When the strip is short, it can look like the letter S.

  DOI: 10.1201/9781003670261-28

WHY IT FORMS A SHAPE WITH TWO CONNECTED SPIRALS

As shown in the photo at the beginning, you might already know from experience that when you peel an apple by twirling it, the peel forms a shape with two connected spirals. But why does it form two spirals connected together, rather than just a single spiral? It's difficult to prove this mathematically, but let's try to explain it using an intuitive image.

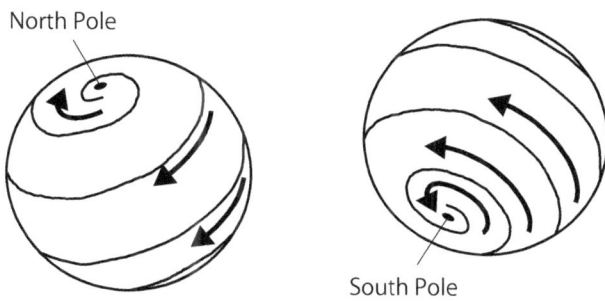

Figure 28.2 Moving clockwise around the north pole.

Imagine starting to peel an apple from the point that corresponds to the North Pole of the Earth, moving in a clockwise direction, and continuing all the way down to the South Pole. The left side of the diagram above shows how a spiral emerges clockwise from the North Pole.

Now, let's flip the apple over and look at it from the South Pole side. As you can see in the diagram on the right, the spiral now appears to turn counterclockwise. What looked clockwise was only from the perspective of the North Pole—when flipped, it appears to spin in the opposite direction.

The same idea applies to the apple peel. In the Northern and Southern Hemispheres, the direction of the spiral appears reversed when viewed from above. So, when you lay the peel flat with both ends—the starting point at the North Pole and the endpoint at the South Pole—facing upward, you'll naturally see two spirals spinning in opposite directions.

One spiral moves from the center outward, and the other from the outside inward. Since they connect in the middle, the result is a shape made of two spirals spinning in opposite directions—or, in some cases, an S-shaped curve.

A NET OF A POLYHEDRON

As another approach, we can represent a sphere as a collection of polygons and observe the net of that shape. This method happens to be my personal favorite. We start by preparing a polyhedron like the one shown in the diagram below.

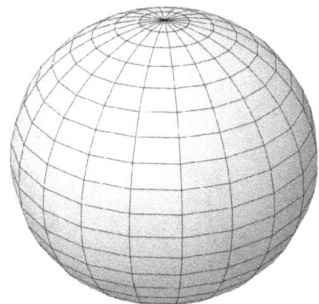

Figure 28.3 A 3D CG model of a polyhedron shaped a sphere.

We place the polygons one by one so that they form a single, continuous net. Starting near the pole, the arcs gradually widen, and around the equator, they become nearly straight. The same pattern appears on the opposite side, so in the end, you can see that the net forms a shape with two spirals spinning in opposite directions, connected together.

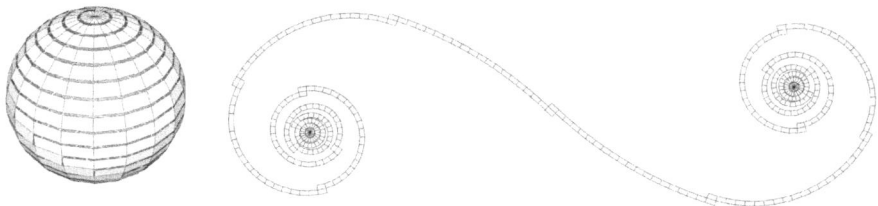

Figure 28.4 A net of a polyhedral model shaped a sphere.

To further recreate the apple-peeling process, I created a 3D model. The curved surface is approximated using a series of long, narrow triangles—this is known as a *triangle strip*.

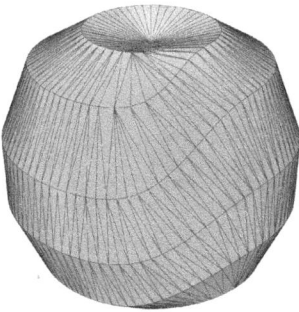

Figure 28.5 A triangular mesh model of an apple.

Using software to create a net,[1] we unfolded the model onto a flat surface. The result beautifully reproduced the appearance of a peeled apple skin.

Figure 28.6 A net of an apple-shaped object.

The relationship between a 3D shape and its net is truly fascinating. If you assemble this net, you can create a papercraft model of an apple.

AN UNUSUAL NET OF A CUBE

Since we've been talking about nets, let me introduce an especially interesting one. Take a look at the diagram below—can you guess what kind of net this is?

---

[1] *Pepakura Designer* — https://pepakura.tamasoft.co.jp/pepakura_designer/ is the best tool for this purpose

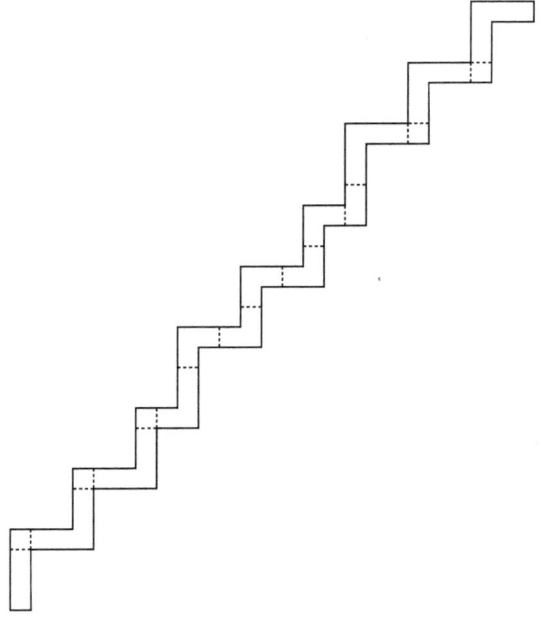

Figure 28.7 A net of what?

Believe it or not, this is actually a net of a cube. By unfolding the cube into a long, narrow strip—much like peeling an apple—you can create a staircase-shaped net like this one.[2] Since it doesn't involve any twirling motion, there are no spirals here.

If you're curious whether it truly forms a cube, try copying this page and assembling the net yourself!

---

[2]Thanks to my college friend Keisuke Nakano for showing me this.

CHAPTER 29

Fractal Figures Drawn by Coloring Grid Cells

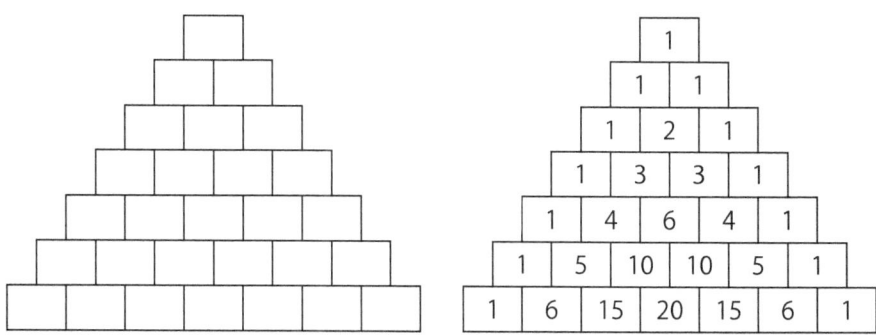

Figure 29.1   Pascal's triangle.

Let's fill in the grid cells arranged in a pyramid, as shown on the left side of the figure, using the following rules:

- Place a 1 in the top row.

- In the rows below, place a 1 in the cells at both ends, and in every other cell place the sum of the numbers directly above to the left and right.

If we keep filling in the numbers row by row according to this rule, we get the arrangement shown on the right side of the figure. This simple rule generates a number pattern known as *Pascal's Triangle*, which has many fascinating mathematical properties.

DOI: 10.1201/9781003670261-29

PASCAL'S TRIANGLE AND BINOMIAL COEFFICIENTS

Let's jump right in and expand the following expressions:

$$(x + y)^1$$

$$(x + y)^2$$

$$(x + y)^3$$

$$(x + y)^4$$

When expanded, they become:

$$(x + y)^1 = x + y$$

$$(x + y)^2 = x^2 + 2xy + y^2$$

$$(x + y)^3 = x^3 + 3x^2y + 3xy^2 + y^3$$

$$(x + y)^4 = x^4 + 4x^3y + 6x^2y^2 + 4xy^3 + y^4$$

Many of you may already have these formulas memorized. Now, let's take the coefficients from the right-hand side of each equation and list them in order from left to right:

$$(x + y)^1 = x + y \quad \rightarrow \quad 1, 1$$

$$(x + y)^2 = x^2 + 2xy + y^2 \quad \rightarrow \quad 1, 2, 1$$

$$(x + y)^3 = x^3 + 3x^2y + 3xy^2 + y^3 \quad \rightarrow \quad 1, 3, 3, 1$$

$$(x + y)^4 = x^4 + 4x^3y + 6x^2y^2 + 4xy^3 + y^4 \quad \rightarrow \quad 1, 4, 6, 4, 1$$

If we compare this with the earlier figure, we see that the sequence of coefficients obtained by expanding $(x + y)^n$ matches exactly with the numbers in Pascal's Triangle. In fact, if we take the very top 1 in the triangle as the 0-th row, then the coefficients of $(x + y)^n$ align perfectly with the numbers in the n-th row.

Once we know this rule, we can immediately figure out the coefficients when expanding $(x + y)^5$. Looking at the 6th row of Pascal's Triangle, we find

$$1, 5, 10, 10, 5, 1.$$

so the expansion is

$$(x + y)^5 = x^5 + 5x^4y + 10x^3y^2 + 10x^2y^3 + 5xy^4 + y^5.$$

In the same way, $(x + y)^6$ can also be expanded.

The numbers arranged like this are called *binomial coefficients*. In general, the k-th number (counting from the left) in the n-th row is given by

$$\frac{n!}{k!(n - k)!}.$$

This expression also represents the number of ways to choose k items from n items. It is often written as

$$_nC_k \quad \text{or} \quad \binom{n}{k}.$$

FIGURES CREATED BY COLORING GRID CELLS

If we color each cell black when the number is odd, and white when the number is even, we get a figure below.

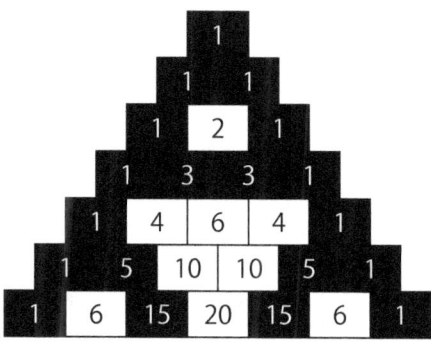

Figure 29.2 The odd numbers in Pascal's triangle are colored in.

The relationships for adding odd and even numbers are as follows:

- Odd + Odd = Even

- Even + Even = Even

- Odd + Even = Odd

Therefore, if the upper-left and upper-right cells are the same color, we paint the current cell white □ (for an even number). Otherwise, we paint it black ■ (for an odd number). Using this method, we can determine the colors without actually performing the addition.

In this way, we can create patterns like the one shown below.

Figure 29.3 Sierpinski gasket.

This figure is also known as the *Sierpiński Gasket*, a type of fractal with self-similarity.

The first part of our discussion may have leaned a bit too heavily on mathematics. But the second part is simply about coloring grid cells according to a very easy rule. If you have some graph paper, you can try it right away.

When using graph paper, unlike the initial figure where the cells were staggered half a step, you should leave one cell of spacing between the colored ones.

The following picture shows a figure drawn by shading the cells of graph paper with a pencil. Give it a try as a quick change of pace—or perhaps as a little escape from reality!

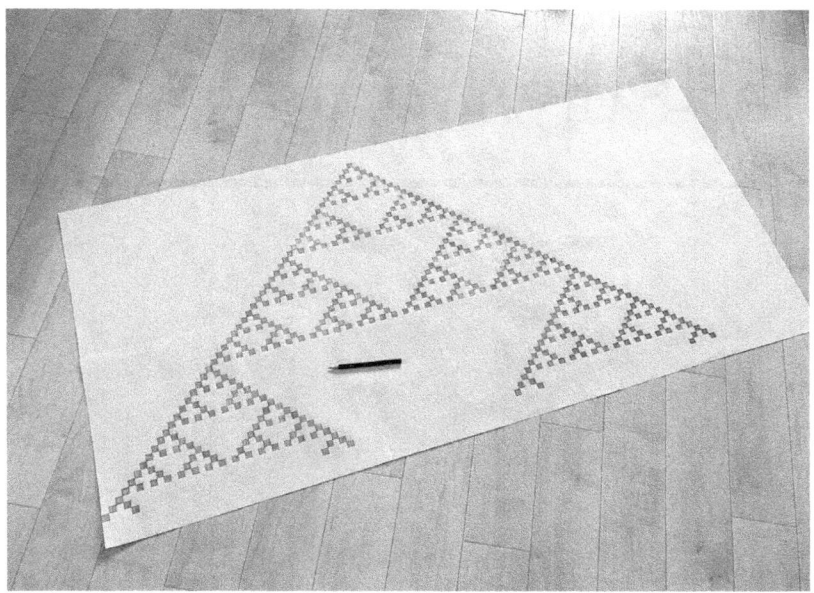

Figure 29.4 Sierpinski gasket drawn by filling in the squares of graph paper with a pencil.

CHAPTER 30

A Rather Casually Made Star Polyhedron

Figure 30.1 Star-shaped polyhedron made by gluing cardboard together.

When you have a lot of leftover cardboard, why not use it for a craft project? It's large, sturdy, and easy to dispose of once you're done. Quite convenient, isn't it?

 DOI: 10.1201/9781003670261-30

If you're not sure what to make, try cutting out many identical isosceles triangles and assembling them into a *star-shaped polyhedron*.

A SHAPE MADE UP OF ISOSCELES TRIANGLES

Once you cut out an isosceles triangle of a suitable size from the cardboard, use it as a template and cut out as many identical triangles as you can. By combining them, you can create a beautiful object like the one shown in the opening figure.

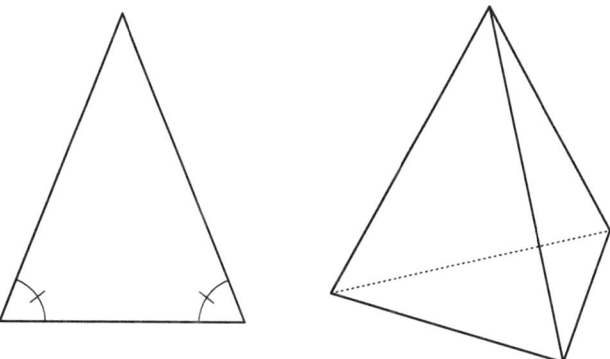

Figure 30.2 Isosceles triangle (left) and equilateral pyramid made by combining three of them.

The dimensions of the base and height of the isosceles triangles can be whatever you like. The important thing is that you prepare many triangles of exactly the same shape. How many will you need? Try making a guess by looking at the opening figure.

The shape in the figure looks like an icosahedron with a triangular pyramid attached to each face. Since there are 20 triangular pyramids, and each pyramid is made of 3 isosceles triangles, the total number required is

$$3 \times 20 = 60.$$

Preparing all the pieces takes some effort, but once you have enough triangles, the rest is just a matter of taping them together with packing tape. First, join three isosceles triangles to form the lateral faces of a triangular pyramid. Since all the triangles are identical, the base will automatically form an equilateral triangle.

After making 20 of these triangular pyramids, the next step is to connect their bases together to form a solid. Imagine attaching each pyramid to the triangular faces of an icosahedron, as shown in the figure below. The key point is to arrange them so that five pyramids meet around each vertex.

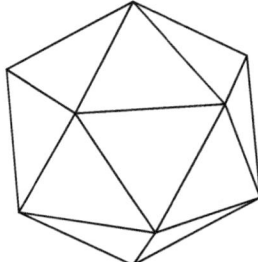

Figure 30.3 Icosahedron.

STAR POLYHEDRON

One example of a shape made by attaching triangular pyramids to a regular polyhedron is the figure shown below, called the *Great Stellated Dodecahedron.*

For it to be called a Great Stellated Dodecahedron, the four points marked with circles in the figure must lie on the same straight line. This happens when the ratio of the base to the side of the isosceles triangle is the *golden ratio,* namely

$$1 : \frac{1 + \sqrt{5}}{2}.$$

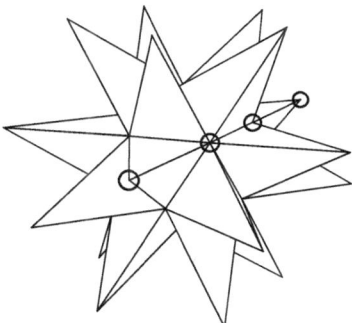

Figure 30.4 Great star dodecahedron.

Another beautiful solid can be made by attaching pentagonal pyramids, each constructed from 5 isosceles triangles, onto the 12 regular pentagonal faces of a dodecahedron, as shown in the next figure. This solid is called the *Small Stellated Dodecahedron*.

Since there are 12 pentagonal pyramids, and each requires 5 isosceles triangles, the total number of triangles needed is

$$5 \times 12 = 60.$$

Interestingly, this also turns out to be 60, just like before.

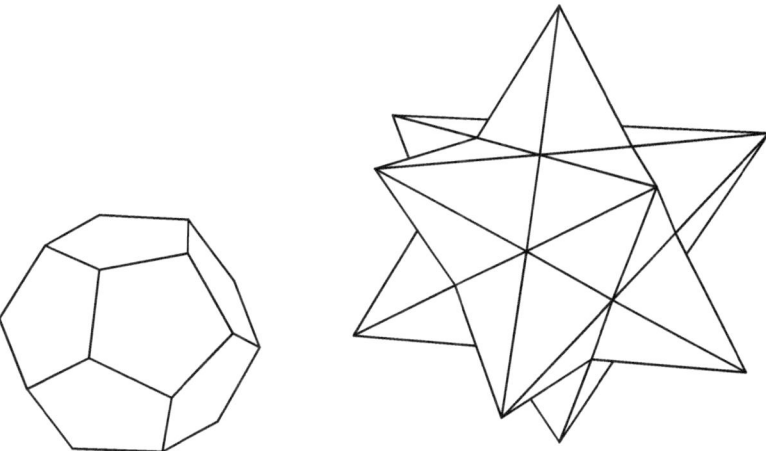

Figure 30.5 Regular dodecahedron and small star dodecahedron.

If you're not aiming to make a perfectly accurate shape, you can simply cut out lots of isosceles triangles and assemble them to easily create a large star-shaped object. The cutting and assembly of the cardboard doesn't need to be very precise. The result will still be a striking object—and the process of making it is enjoyable in itself.

CHAPTER 31

The Wonder of Origami Crease Patterns

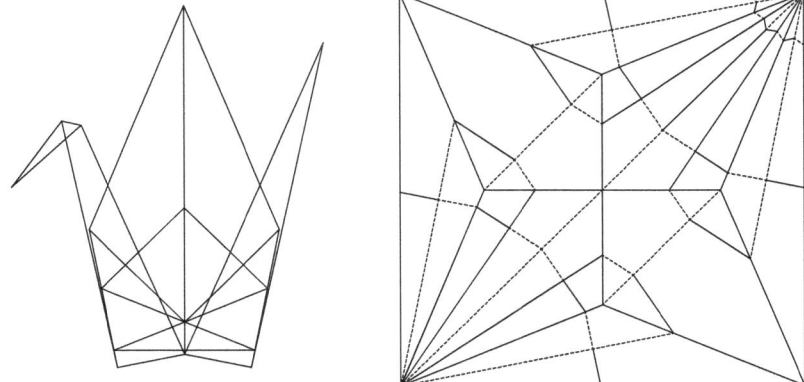

Figure 31.1   The crease pattern of an origami crane.

Origami allows us to create a wide variety of shapes, but most of these designs are completed by repeatedly folding the paper flat. Even models that appear three-dimensional are typically folded flat up until the final steps. For example, in the traditional Japanese origami crane, all steps except for spreading the wings involve flat folds. In this way, folding paper flat is a fundamental operation in origami, and this process is called *flat-folding*.

The creases formed through flat-folding are always straight lines. As a result, when we unfold a flat origami model, the resulting crease

DOI: 10.1201/9781003670261-31

pattern is a collection of straight lines. The illustration at the beginning shows the shape of a crane just before its wings are spread (with the crease lines and edges of the paper shown as if seen through the paper), along with its corresponding crease pattern.

A crease pattern can be understood as the record of the creases left on the paper after folding. Therefore, to recreate the crane, one simply needs to fold the paper along these crease lines.

COLORING POLYGONS IN A CREASE PATTERN

A crease pattern is created by unfolding the paper after it has been folded, and extracting only the creases that were actually used to form the shape. It does not include any creases that were used merely as reference marks and did not contribute to the final model.

Crease patterns contain polygons formed by the surrounding crease lines. There are two types of creases: *mountain folds* and *valley folds*. These are commonly distinguished using dash-dot lines and dashed lines,[1] but in this book, we will use solid lines and dashed lines, respectively.

Regardless of whether a fold is a mountain or a valley, folding along a crease involves bending the paper by 180°. As shown in the following diagram, one of the two polygons adjacent along a crease will end up with the back side of the paper (the white side in origami) facing upward.

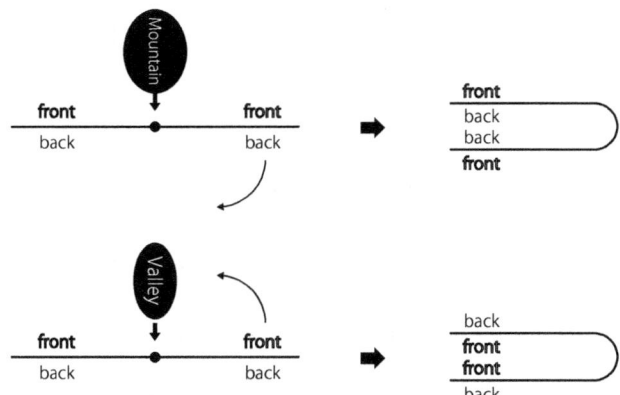

Figure 31.2 The relationship between paper folds and the front and back sides of the paper.

---

[1]When using colors to distinguish between them, mountain folds will be shown in red and valley folds in blue.

If we use white to represent polygons with the back side of the paper facing up, and gray for those with the front side facing up when folded, then the crease pattern from the opening diagram would look like the one below. Along each crease line, one side is white and the other is gray.

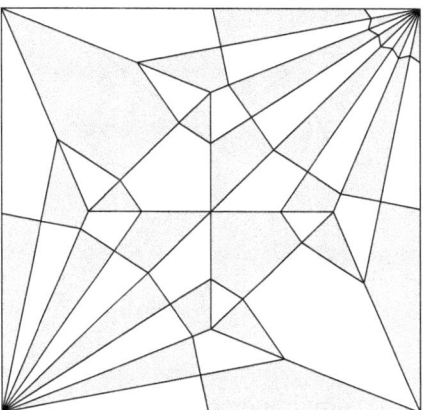

Figure 31.3 Coloring the crease pattern of an origami crane.

Since the two polygons adjacent across a crease line never have the same color, crease patterns in origami have an interesting property: *they can always be colored using just two colors* (In contrast, it is known that general maps require up to four colors to ensure that no adjacent regions share the same color.[2])

Let's try unfolding a folded origami model and coloring the polygonal regions using two colors.

Also, for a map to be colorable with just two colors, the number of lines (or creases) meeting at any point must be even. This fact is connected to a theorem we'll explore shortly.

CHARACTERISTICS OF CREASES AROUND A VERTEX

Let's continue exploring the crease pattern. If we focus on the points where the ends of crease lines meet—these are called *vertices*—we can

---

[2]This is widely known as the *Four-Color Problem*. It remained unproven for over 120 years until it was finally resolved in the 20th century by Kenneth Appel and Wolfgang Haken, becoming the *Four-Color Theorem*.

isolate the area around one of them. One example of such a region is shown in the diagram below.

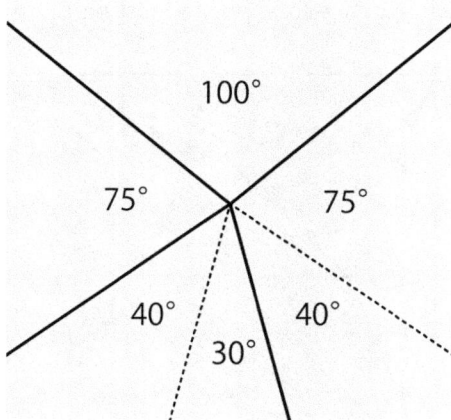

Figure 31.4 Flat-foldable single-vertex crease patterns.

A crease pattern with only one vertex like this is called a *single-vertex crease pattern*. When such a pattern is known to be flat-foldable, the arrangement of creases around the vertex always satisfies the following property:

The sum of alternating angles around the vertex is 180°.

If you take the angles from the previous diagram and add every other one, starting for example with the 100° angle, you get $100° + 40° + 40° = 180°$, which confirms this property. Interestingly, this always holds true for any flat-foldable single-vertex crease pattern, regardless of its specific shape. This result is known as *Kawasaki's Theorem*.[3]

In addition, when we look at the number of mountain and valley folds, the following relationship also holds:

(Number of mountain folds) − (Number of valley folds) = ±2.

Referring again to the previous diagram, there are 4 mountain folds and 2 valley folds, so the equation holds. This result is called *Maekawa's*

---

[3] Also known as the Kawasaki–Justin Theorem.

Theorem.[4] Furthermore, the total number of mountain and valley folds is always even. This matches the property mentioned earlier for two-colorable maps: the number of lines (or creases) meeting at a point must be even.

This property always holds for any crease pattern that can be folded flat. Try folding a piece of paper yourself, then unfold it and observe the pattern.

WHICH CREASE PATTERNS CAN BE FOLDED?

Earlier, Kawasaki's Theorem and Maekawa's Theorem described properties of crease patterns that *can* be folded flat. But what about the reverse? If a crease pattern satisfies these two conditions, can we say that it *must* be flat-foldable?

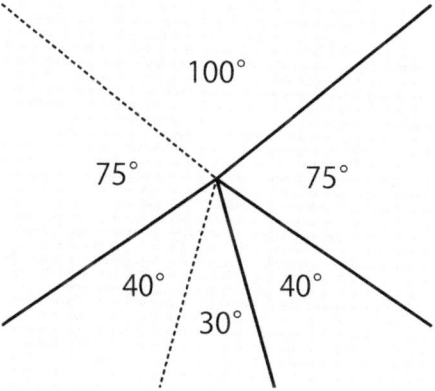

Figure 31.5 Single-vertex crease patterns that cannot be flat-folded.

The crease pattern in the diagram above looks almost identical to the previous one. The arrangement of crease lines is exactly the same, with only a slight difference in the assignment of mountain and valley folds. The alternating angles still sum to 180°, and the difference between the number of mountain and valley folds is 2. At first glance, it seems like it should fold just fine. However, in reality, this pattern *cannot* be folded flat.

---

[4] Also known as the Maekawa–Justin Theorem.

It's quite difficult to explain in words why it doesn't fold, so try it out with a piece of paper. You'll experience for yourself how the paper ends up colliding with itself, making it impossible to complete the fold.

Since the arrangement of crease lines is identical in both crease patterns, we can see that the problem lies in the assignment of mountain and valley folds. It is known that if the crease pattern satisfies Kawasaki's Theorem, then there always exists a way to assign mountain and valley folds that allows the pattern to be folded flat.

Putting all this together:

- Satisfying Kawasaki's Theorem is a *necessary and sufficient condition* for a single-vertex crease pattern to be flat-foldable.

- Satisfying Maekawa's Theorem is a *necessary condition*, but not a sufficient one.

That might sound a bit technical, so let's take a moment to explain what these terms mean: *necessary condition*, *sufficient condition*, and *necessary and sufficient condition*—and go over them carefully to make sure the idea is clear.

Necessary Condition
When condition B *must* be true in order for condition A to be true, we say that B is a *necessary condition* for A. However, even if B is true, that doesn't mean A is always true.

> **Example:**
> To say '(Condition A) Person Y lives in Tokyo," it must also be true that '(Condition B) Y lives in Japan." So, B is a necessary condition for A.

> **Origami Example:**
> For a single-vertex crease pattern to be flat-foldable, it *must* satisfy Maekawa's Theorem. However, just because it satisfies Maekawa's Theorem doesn't mean it can always be folded flat.

Sufficient Condition
When condition A being true *guarantees* that condition B is also true, we say that A is a *sufficient condition* for B.

Example: If "(Condition A) Person Y lives in Tokyo" is true, then "(Condition B) Y lives in Japan" is definitely true. So, A is a sufficient condition for B.

Origami Example:

If a single-vertex crease pattern can be folded flat, then it always satisfies Kawasaki's Theorem. And if a crease pattern satisfies Kawasaki's Theorem, then (with an appropriate mountain-valley assignment) it *can* be folded flat.

Necessary and Sufficient Condition

When A being true always implies B is true, *and* B being true always implies A is true, then B is a *necessary and sufficient condition* for A. In other words, A and B are logically equivalent.

Example:

"(Condition A) X is even" is true if and only if "(Condition B) The square of X is even" is also true for a positive integer X.

Origami Example:

A single-vertex crease pattern is flat-foldable *if and only if* it satisfies Kawasaki's Theorem. This makes Kawasaki's Theorem a necessary and sufficient condition for flat-foldability.

Terms like "necessary condition" and "sufficient condition" might not come up often in everyday life, but thinking about flat-foldability in origami is a great way to review logical reasoning—something you learn in high school math. It's a useful way to practice thinking logically about a variety of topics.

By the way, everything we've discussed so far has been about crease patterns with just one vertex. When a pattern has multiple vertices, determining whether it can be folded flat becomes a much more complex problem.[5] Even a single origami crease pattern can contain a surprising amount of depth!

---

[5] The difficulty of determining whether a given crease pattern can be folded flat belongs to a class of problems known as *NP-hard*.

CHAPTER 32

How Can We Fit As Much As Possible into an Envelope?

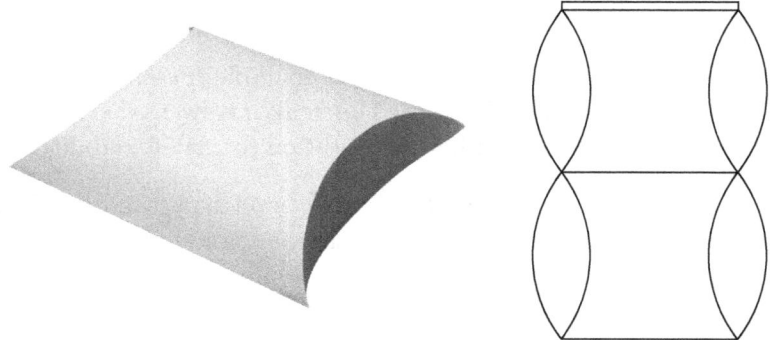

Figure 32.1   Pillow box and its net diagram.

Packages made of paper come in a variety of shapes, but it's rare to see ones that incorporate curves. Among the few that do, a well-known example is the *pillow box*.

A pillow box has the shape shown on the left side of the figure above, and its template, shown on the right, features curved fold lines. These boxes are widely used for packaging all sorts of items, from clothing to food. In fact, McDonald's apple pie packaging uses a similar design.

                  DOI: 10.1201/9781003670261-32

When the paper is folded along the curves in the template, it forms a smooth, rounded surface.

So, can these fold lines be shaped however we like? Or is there a specific shape they must follow?

SHAPE OF THE FOLD LINES IN A PILLOW BOX

When we think about the shape of the fold lines, it's clear that the curve on the right side of the figure below won't work. On the other hand, the one on the left seems fine. But what about the one in the middle? Do you think it would work? Or not?

This raises an interesting question: what are the necessary conditions for a curve to work as part of a pillow box design?

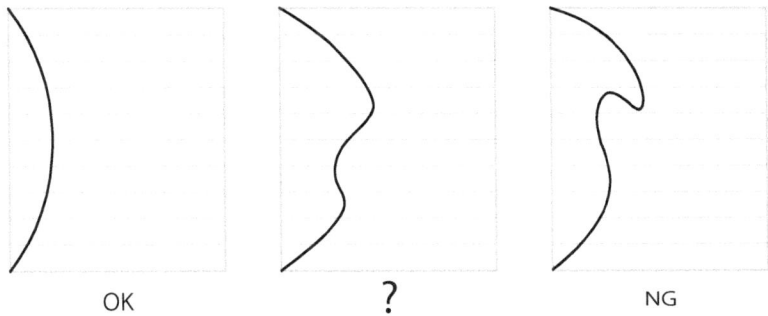

OK ? NG

Figure 32.2 Which curves can be used for the fold lines of a pillow box?

Let's clearly define the type of pillow box shape we're focusing on. Using the terminology shown in the figure below, we'll consider only those that meet the following conditions:

- The shape is symmetric both left-to-right and top-to-bottom (the figure shows one-quarter of the whole shape).

- Both the top and the side surfaces are cylindrical. The top consists of a collection of horizontal straight lines, while the side consists of vertical straight lines.

- The box can be made from a rectangular sheet of paper, forming a fully closed shape with no gaps.

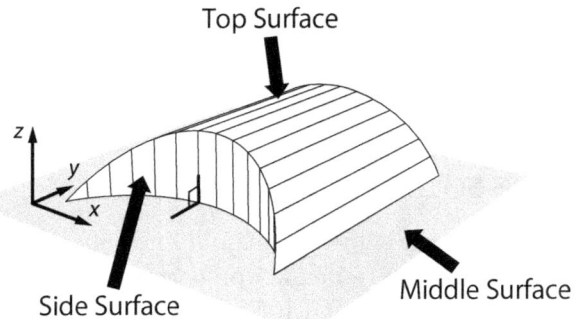

Figure 32.3 Target pillow box shape (one-fourth of the whole).

As shown on the left side of the figure below, if we approximate the surface using a collection of narrow quadrilateral planes (this is called a *discrete representation* of the curved surface), then due to geometric constraints required for the triangle $\triangle SPQ$ in the diagram to exist, we find that the angle θ must satisfy the condition $-\pi/4 \leq \theta \leq \pi/4$.

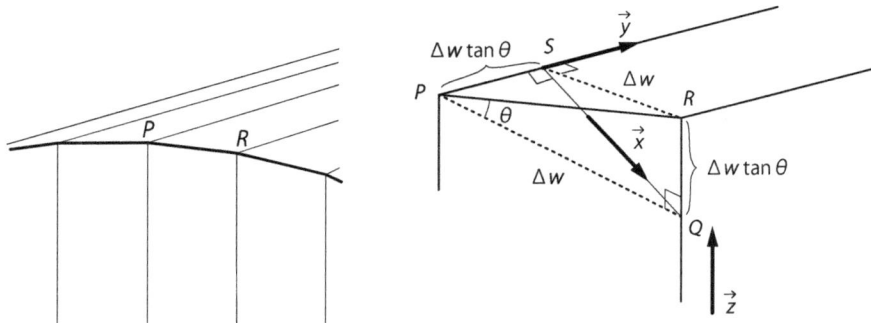

Figure 32.4 Close-up of the fold area.

This can be derived as follows from the Pythagorean Theorem.

$$|PS|^2 + |SQ|^2 = |PQ|^2$$

$$\Rightarrow (\Delta w \tan \theta)^2 + |SQ|^2 = \Delta w^2$$

$$\Rightarrow (\tan \theta)^2 = 1 - \frac{|SQ|^2}{\Delta w^2}$$

$$\Rightarrow \ |\tan\theta| \leq 1$$

$$\Rightarrow \ -\pi/4 \leq \theta \leq \pi/4.$$

Δw is the distance between the vertical fold lines passing through points P and R, and θ is the angle $\angle QPR$, which corresponds to the slope of the fold line in the unfolded template (see the figure below).

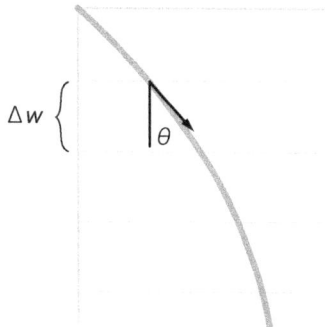

Figure 32.5   Slope of the tangent line to the fold curve.

From this analysis, we can conclude that *a valid fold line curve must have a tangent whose slope stays within ±45 degrees.*

If this is the only condition that needs to be met, then we actually have a lot of freedom in choosing the shape of the fold lines. To test this, I tried creating different types of pillow boxes. On the left side of the figure below, I used a zigzag line, and on the right, a smooth curve.

In both cases, I was able to successfully create pillow boxes that are a bit more stylish and unique than the usual design.

I was also able to create a variety of other designs, as shown in the figure below.

While some of these designs may not be practical in terms of ease of assembly, strength, or internal volume, I was still able to create pillow boxes with a completely different look from the traditional style.

Moreover, if we relax the conditions of top-bottom symmetry and vertical side surfaces, we can even make designs like the ones shown in the figure below. It seems that pillow boxes hold a lot of potential for creative design.

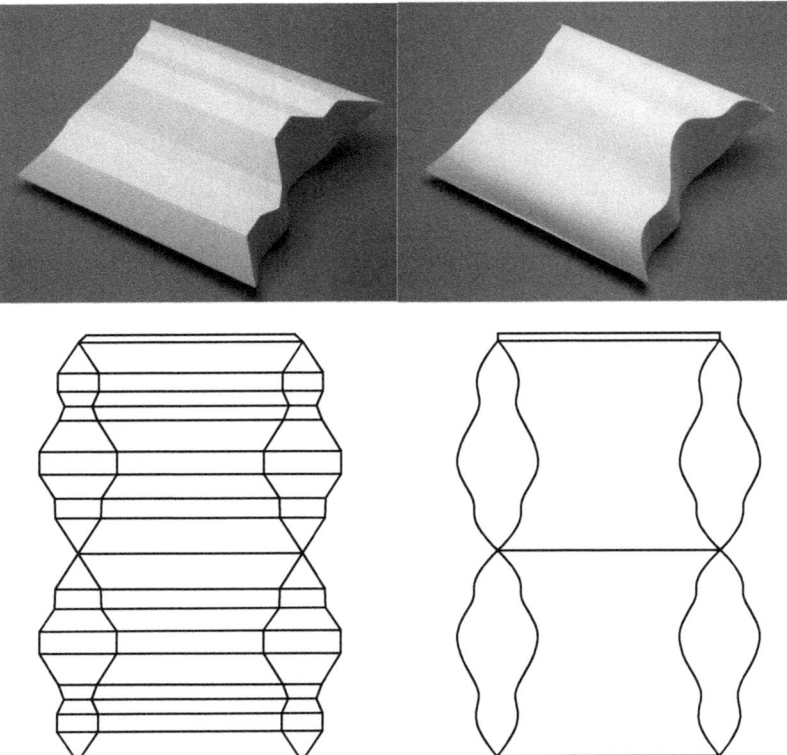

Figure 32.6 New pillow box design.

VOLUME OF A 3D OBJECT THAT FITS IN A LETTER PACK

One of the services offered by Japan Post is called "Letter Pack Plus." With this service, as long as your item fits into a designated envelope, you can send it anywhere in Japan for a flat rate.[1]

A key point of this service is that there's no restriction on the thickness of the envelope after it has been filled. In other words, it's perfectly fine to fold the envelope into a 3D shape. Given the flat rate, it's only natural to want to fit in as much as possible. So, what if we fold the envelope into a pillow box shape? How much volume would it have? And what kind of fold line shape would maximize that volume?

---

[1]There is a weight limit of 4 kg.

Figure 32.7 Various pillow box designs.

This type of question—finding the maximum value of something under certain conditions—is known as an *optimization problem*. Here, let's
consider the following specific problem:

Question: If we fold an envelope of Letter Pack into the shape of a
pillow box, what fold line curve will maximize its volume? And what is
the maximum volume in that case?

The optimal shape of the curve depends on the aspect ratio of the
original envelope. For this calculation, we used the actual dimensions of
a Letter Pack: 340 mm by 248 mm (giving a vertical-to-horizontal ratio
of 1.37, which is slightly different from the 1.41 ratio of standard copy
paper).

We'll assume the fold line curve is symmetric, and consider only
one side of it. That side will be limited to a *cubic Bézier curve*.
A cubic Bézier curve is a type of curve commonly used in design
software like Adobe Illustrator, and its shape is determined by four
control points. Therefore, to find the curve that maximizes the volume, we need only determine the optimal positions of these control
points.

After running the calculations on a computer, it was confirmed that
the volume reaches its maximum—4,309 cm³—when the fold line takes

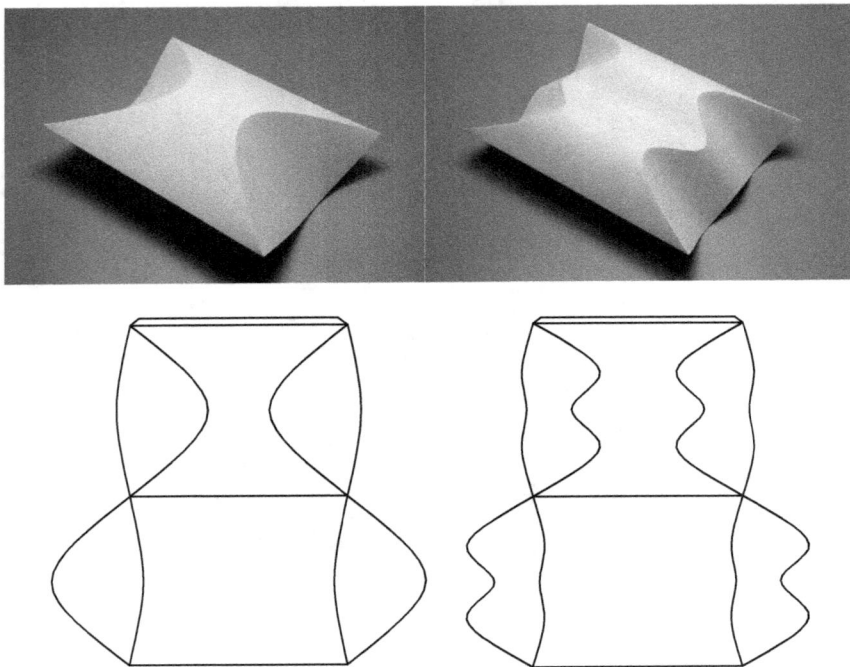

Figure 32.8 Asymmetrical pillow box design (top and bottom not aligned).

the shape shown in the figure below. That means you can fit the equivalent of over 4 liters into an envelope that's just slightly larger than A4 size.

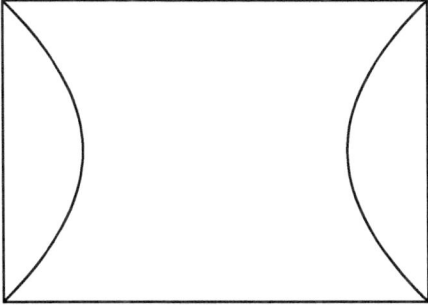

Figure 32.9 Net diagram of the pillow box with maximum volume (one side only).

To compare with the volume of the pillow box, I also calculated the maximum volume for cases where the cross-section is a cylinder, a rectangle, or a rhombus. The results are shown in the figure below.

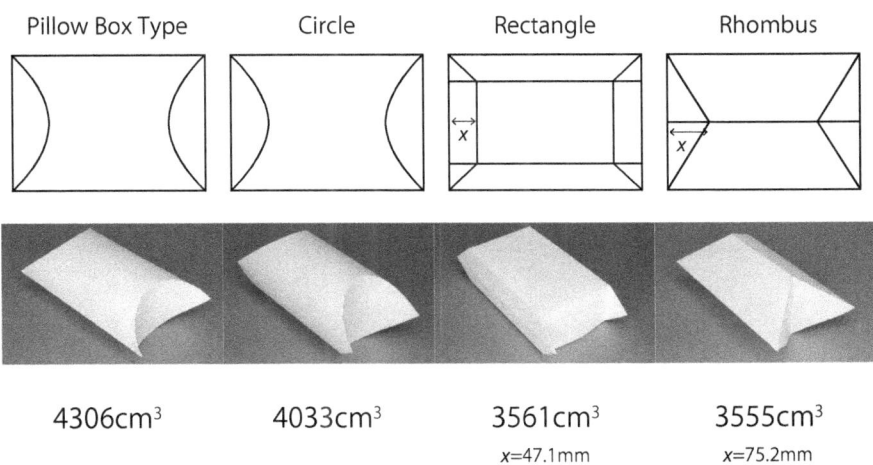

Pillow Box Type	Circle	Rectangle	Rhombus
4306cm³	4033cm³	3561cm³	3555cm³
		x=47.1mm	x=75.2mm

Figure 32.10 Relationship between folding method and volume.

Interestingly, it turned out that the pillow box shape provides the largest volume among the options. This might be one of the reasons why the pillow box is so widely used.

It seems that the typical pillow box design is not only appealing in terms of aesthetics, but also quite effective when it comes to maximizing volume.

CHAPTER 33

Spiral Made from Paper Tape

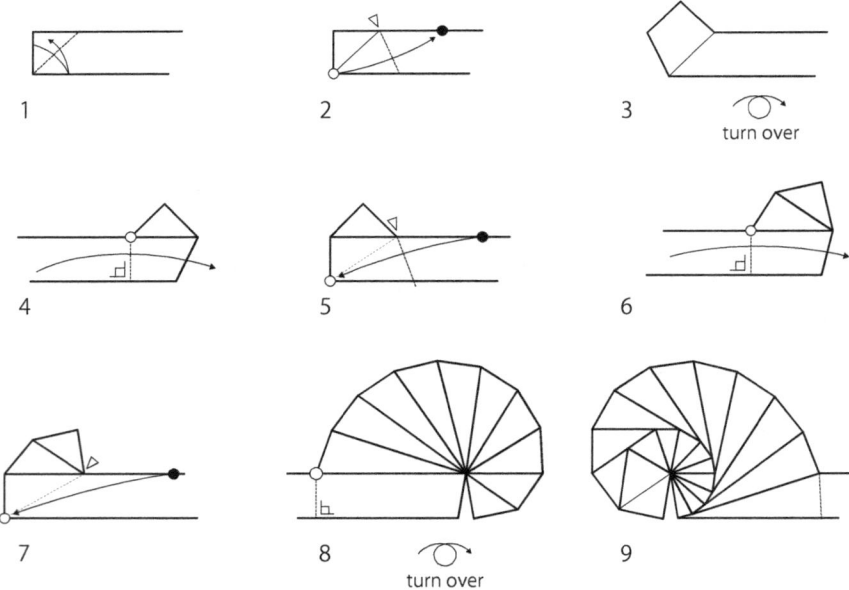

Figure 33.1 Steps for folding a spiral with paper tape.

By following the steps shown in the figure above, you can fold a strip of paper into a beautiful, shell-like spiral shape. This is a method for

DOI: 10.1201/9781003670261-33

creating spirals with paper tape, which I learned at a workshop by origami artist Tomoko Fuse.[1]

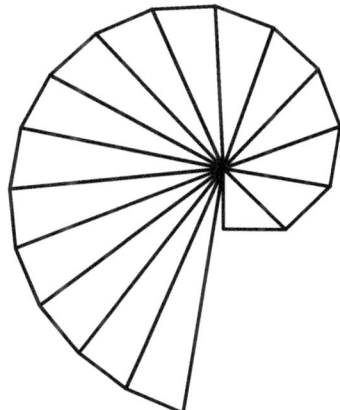

Figure 33.2 Spiral shape made from paper tape.

By simply folding a strip of paper according to a few basic rules—without using any tools—you can create a spiral shape like the one shown in the figure above. The only folding actions required are folding so that a corner lands on the edge of the paper tape, and folding back along a line perpendicular to the edge.

It's surprising that such an elegant spiral shape can be created with just these simple steps.

TYPES OF SPIRALS

So, what kind of spiral is formed by this folding method?

There are many types of spirals, but two well-known ones are the *logarithmic spiral* and the *algebraic spiral* (also known as the *Archimedean spiral*). As shown in the figure below, a logarithmic spiral resembles a seashell, with the spacing between the arms increasing as it moves outward from the center. This type of spiral is commonly seen in nature.

In contrast, an algebraic spiral maintains a constant spacing between the arms.

---

[1]Reference: Fuse Tomoko, SPIRAL FROM TAPE, SPIRAL–ORIGAMI — ART — DESIGN, VIRECK VERLAG, p.51, 2012

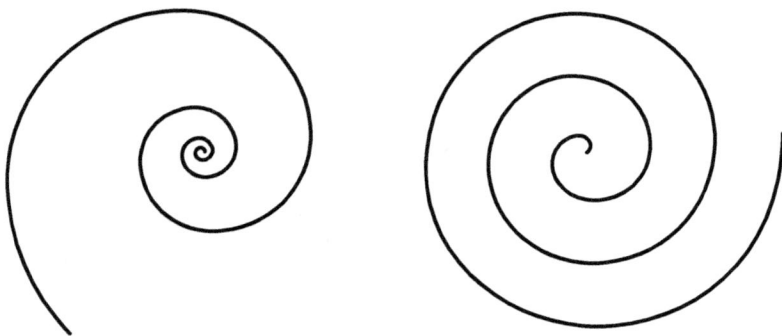

Figure 33.3 Logarithmic spiral (left) and algebraic spiral (right).

For a point on a spiral, the relationship between its distance from the origin r and its angle of rotation θ (called the *polar angle*) can be expressed with the following equations (here, B and a are constants):
Logarithmic spiral:

$$r = B^{\theta}$$

Algebraic (Archimedean) spiral:

$$r = a\theta$$

Expressing a point's position based on its distance from the origin and its angle in this way is known as a *polar coordinate representation*.

On a logarithmic spiral, the distance from the origin increases exponentially with the angle θ. In contrast, on an algebraic spiral, the distance increases linearly with θ.

At first glance, the spiral formed by folding the paper tape appears similar to a logarithmic spiral. However, as the folding continues, its behavior begins to differ.

The figure below shows the spiral that emerges after 300 folding steps. It appears to be an algebraic spiral, with a constant spacing between the arms. How curious!

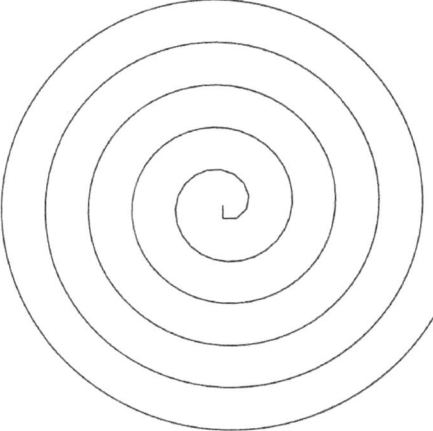

Figure 33.4 The spiral shape obtained by folding paper tape up to 300 times.

CHECK BY CALCULATION

Let's set the width of the paper tape to 1. The spiral formed by folding the tape can be represented as a sequence of triangles, so let's examine the lengths of their sides.

First, the initial triangle turns out to be a right isosceles triangle, with both of its equal sides equal to the width of the tape. Let's denote the lengths of the sides extending outward from the center as $a_1, a_2, a_3, \ldots, a_n$.

The first length is $a_1 = 1$, and the next is $a_2 = \sqrt{2}$. Continuing in this way, the next length a_3 can be found using the Pythagorean theorem: since $a_3^2 = 1 + a_2^2 = 3$, we get $a_3 = \sqrt{3}$.

In this way, we can see that the general expression is $a_n = \sqrt{n}$.

It turns out there's a remarkably clear pattern behind this spiral. This type of spiral is known as the *Theodorus spiral*. The Theodorus spiral looks like the one shown in the figure below.

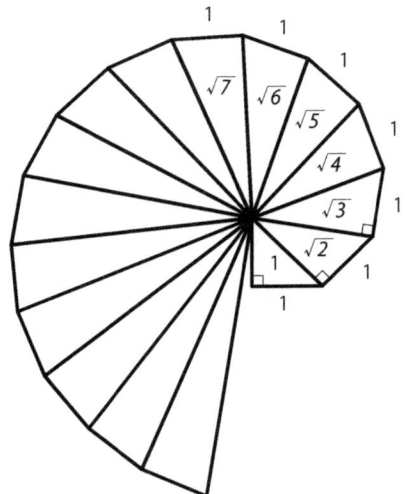

Figure 33.5 Theodolite spiral.

Let's now take a look at how the shape of the curve changes as we increase the number of folds.

When a new vertex is added by folding, we can represent it in polar coordinates. The increase in distance dr from the origin, when moving from the nth point to the $(n + 1)$th point, is given by:

$$dr = a_{n+1} - a_n = \sqrt{n + 1} - \sqrt{n}.$$

At the same time, the increase in the angular coordinate $d\theta$ is

$$d\theta = \arccos\left(\frac{a_n}{a_{n+1}}\right) = \arccos\left(\frac{\sqrt{n}}{\sqrt{n + 1}}\right).$$

The function arccos is the inverse of the cosine function. When there is a relationship $x = \cos(\theta)$, the angle θ can be expressed as:

$$\theta = \arccos(x).$$

Using mathematical software, we can compute the limit of $\frac{dr}{d\theta}$ as n becomes large—that is, the rate of change of r with respect to the polar angle θ. The result is

$$\lim_{n \to \infty} \frac{dr}{d\theta} = \frac{1}{2}.$$

This means that as we continue folding and n grows large, the increase in distance r from the origin approaches half the increase in the polar angle θ. Since one full turn of the spiral increases θ by 2π, the corresponding increase in r will be π.

From this, we see that the shape of the spiral gradually approaches that of a spiral with a constant width of π, as shown in the figure below.

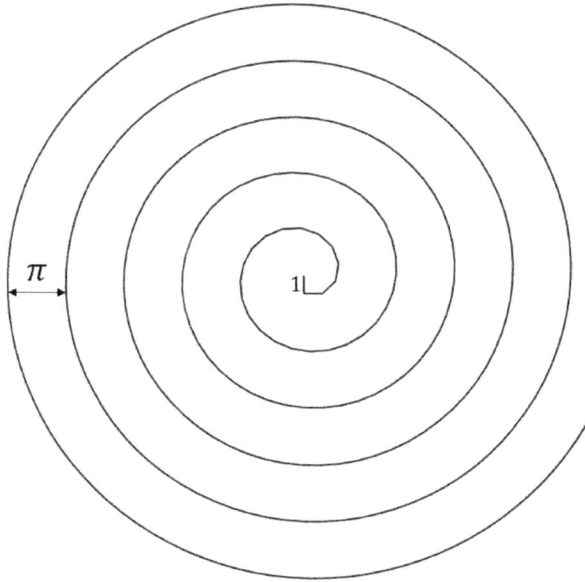

Figure 33.6 If you keep folding it, the width of the spiral will approach Pi.

It's fascinating that by simply folding a spiral with a strip of paper, the number π naturally appears in the process.

Afterword

Ever since I was a child, I loved making things—I would often cut up paper to create all sorts of crafts. In elementary school, I was especially fond of math, particularly geometry and puzzles that revealed elegant solutions when approached from the right angle. After receiving a computer from my father, I became obsessed with programming and drawing various shapes on the screen.

Later, I read "Aha! Gotcha" by Martin Gardner, the founding figure of recreational mathematics, and was deeply inspired. I also found myself admiring the brilliant stories in Richard Feynman's "Surely You're Joking, Mr. Feynman!", which sparked a longing in me to become someone intellectually curious like him. Another major influence was Professor Kenjiro Suzuki of the University of Tokyo, whose passionate lectures on the beauty of geometry left a strong impression on me. Encouraged by these mentors and role models, I began to dream of following in their footsteps.

After becoming a university professor, I began occasionally posting about mathematical topics found in everyday life on my personal blog and on X (formerly Twitter). When writing this book, I selected some of the most engaging and accessible topics from those past posts and expanded them to include details that wouldn't fit in a tweet. Initially, I planned to include about 50 or 60 topics, but as I kept adding more and more, the page count ballooned, and I eventually narrowed it down to 33 topics. If readers enjoy this book, I hope to find another opportunity to share the rest.

Among the 33 topics in this book, many relate to paper folding, which is one of my research areas, as well as to Plarail—the toy train system for which I served as a 65th anniversary ambassador—and to hands-on crafting. Working with your hands to build things involves many elements of geometric learning. Looking over the entire book, I also noticed an unexpectedly frequent appearance of logarithms (log). High school math may seem dry at first glance, but this book shows that it's deeply embedded in our everyday lives.

This book became a reality thanks to Ren Munakata at Yama-kei Publishers, who reached out to me with the idea of turning my math-related posts on X into a book. I'm also grateful to her for her valuable suggestions about the book's structure and her support throughout the process. The charming illustrations were provided by Koji Yoshiike. Since the book contains so many diagrams, I'm sure this was a demanding task, and I'm truly thankful for his contribution.

I also want to thank Kiyoshi Kotani and Jun Maekawa for reviewing the manuscript. Kiyoshi and I were classmates at the University of Tokyo, where we studied precision mechanical engineering. Jun Maekawa is a fellow member of the Japan Origami Academic Society and has deep expertise in both origami and mathematics. Their feedback helped make this book even better.

Many of the ideas for the math-in-everyday-life topics featured in this book came from playing and talking with my three children. I'd like to express my heartfelt thanks to my family, who continually provide me with fresh perspectives and creative ideas.

If this book helps readers discover the joy of everyday mathematics, then nothing would make me happier.

Index

For Product Safety Concerns and Information please contact our EU
representative GPSR@taylorandfrancis.com
Taylor & Francis Verlag GmbH, Kaufingerstraße 24, 80331 München, Germany

www.ingramcontent.com/pod-product-compliance
Lightning Source LLC
Chambersburg PA
CBHW070229180526
45158CB00001BA/199